HUMAN ADAP

EDITED BY

G.A. HARRISON

Emeritus Professor of Biological Anthropology
University of Oxford and Emeritus Fellow of Linacre College

HOWARD MORPHY

Professor of Anthropology, University College London;
Professor of Anthropology, Australian National University; and
Honorary Curator, Pitt Rivers Research Centre

BERG

Oxford • New York

First published in 1993 by Oxford University Press
Reprinted with a new Preface in 1998 by
Berg
Editorial offices:
150 Cowley Road, Oxford, OX4 1JJ, UK
70 Washington Square South, New York, NY 10012, USA

Berg is the imprint of Oxford International Publishers Ltd.

Library of Congress Cataloging-in-Publication Data

A catalogue record for this book is available from the Library of Congress.

British Library Cataloguing-in-Publication Data

A catalogue record for this book is available from the British Library.

ISBN 1 85973 958 X (Paper)

Printed in the United Kingdom by WBC Book Manufacturers, Mid-Glamorgan.

HUMAN ADAPTATION

PREFACE

Howard Morphy and Geoffrey Harrison

When visiting a doctor for a general examination one of us was subjected to the usual interrogation on exercise. Playing cricket and squash fairly regularly could perhaps have provided a satisfactory answer. But no, jogging was what was required at the time: 'Playing cricket is not like being chased by a sabre-toothed tiger. For 99 per cent of our history we have been hunters and gatherers, survival of the fittest you know.' There was no point in taking the discussion further by questioning the underlying assumptions about hunter-gatherers, providing more details about Pleistocene environments, or suggesting that squash might indeed have been as good a preparation as any for being chased by a sabre-toothed tiger! The cure was a fixed one, partly fashion, partly medical common sense; the packaging reflected folk wisdom with a little anthropological knowledge filtering through.

The idea of adaptation is a part of Western folk wisdom about the world as well as being integral to Darwinian evolutionary theory. As Malcolm Smith argues in Chapter 1, adaptation is a 'common sense default assumption'. It is very easy to slip between scientific and everyday uses of terms such as 'fitness' and 'adaptation'. It is also easy to confuse adaptation in a causal sense – that the presence of a particular gene or pattern of behaviour is caused by its particular function – with the human capacity to adapt to a particular circumstance. Adaptability, in the sense of the capacity to act flexibly and to respond to very different environmental circumstances, to different social contexts and to different technological problems, is a characteristic of the human species. But to say that a particular response, a particular demonstration of the adaptability of human beings, is the *result* of adaptation is to confuse the capacity to do something with its functional purpose and origin. Of course the two are related, since if individuals show such adaptability then the capacity is a species characteristic produced as a species adaptation by evolution!

The physiological capacity that enables human beings to adapt to living in high altitudes provides an excellent example of the need to separate adaptation as a causal factor in the particular context from the ultimate ways in which the particular adaptation may be exploited. Geoffrey Harrison shows in Chapter 2 how the capacity to adapt to living at high altitude is unlikely to have been caused by the requirements of living in high altitudes, since hominids almost certainly evolved the necessary capacities while living in lowland savannahs. Human beings were preadapted, fortuitously, to living at high altitudes. This may have conferred on them an advantage as a species, but living at high altitudes did not cause the adaptation in the first instance.

Adaptation is an important concept for anthropologists to address since it exists at the intersection between human biology, cultural process and social organization. In this book we intend to view the concept from four different perspectives within anthropology: genetic (Chapter 1), physiological (Chapter 2), behavioural (Chapter 3), and cultural (Chapter 4). In each case we will draw attention both to the importance of the concept of adaptation and to the complexities of applying it. We will show ways in which the concept can be applied in social and cultural as well as biological anthropology, and draw attention to the potential contribution that anthropology, as a holistic discipline, can make to human evolutionary biology.

In evolutionary biology adaptation is a central concept which must apply as much to *Homo sapiens* as to any other species. As a product of natural selection it is most aptly applied to individuals in the sense that one individual is better or worse adapted than others to the environment in which he or she is living, but it can also be extended by kin selection to families and further related groups. Those individuals who have greatest reproductive success due to the advantage given them by their genetic inheritance differentially influence the gene pool in subsequent generations. Over the generations this process has an effect on the kind of creatures human beings are, resulting in an accumulative process of 'advantageous' adaptation. However, it is important to note that adaptation is not advantageous in some absolute sense of the population being better adapted in the long term as a species than it was previously. Rather, the genetic endowment of particular individuals has been advantaged over others in the evolutionary process. It may be that, as a consequence, the population as a whole is better able to utilize the resources of the environment or occupy new ecological niches, but this is not necessarily the case.

The big problem with adaptation in the evolutionary sense is that where it is obvious, at the species level, it cannot be measured because there is no control. On the other hand, where in principle it could be measured, at the individual level, it is far less obvious! In essence evolutionary adaptation occurs through the replacement of one gene by another as a result of natural selection, but natural selection is very difficult to detect and there are other causes of gene replacement. These issues are discussed in detail by Malcolm Smith in Chapter 1.

These problems of scale mean that it is often difficult to demonstrate specific instances of adaptation operating as a factor in genetic change. They also go some of the way to explaining the difficulties of communication that sometimes arise between social and biological anthropologists. It is often difficult for social anthropologists to see how long-term biological processes can influence the minutiae of social life that they observe. While there is growing acceptance among social anthropologists that cognitive factors influence the ways in which people act in the world there is a reluctance to acknowledge that human cognitive processes have been subject to evolutionary processes. Human biology and physiology are likely to be as important as cognitive factors in developing an understanding of many areas of human behaviour and culture. Theoretical topics in social and cultural anthropology, such as the study of the emotions, or the idea that much behaviour is in some senses embodied, can be usefully approached from perspectives that take relevant biological factors into account.

It has been equally difficult for biological anthropologists to understand the relevance of debates in, say, the anthropology of religion to human evolutionary biology (but see Reynolds and Tanner 1995). And yet if religious practice can be such a significant factor in group identity that it influences patterns of marriage it clearly can, in particular cases, affect selection processes. As a general principle it is important to understand the biology of humans as social and culture-bearing animals, and equally important to place social and cultural behaviour in the context of human biological evolution. Many problems in social anthropology can be approached without considering biological factors and vice versa, but those problems can also be viewed within broader and more encompassing frameworks. There is no reason why all anthropologists should be interested in the full compass of the discipline, but it is often that very width which enables anthropology to make its most significant contributions to contemporary debates. The biological anthropologist needs the data on kinship, marriage, paternity and maternity in order to understand and document the local processes by which genetic change occurs. Studies of the long-term development of regional systems require a consideration of the interaction between social, cultural, biological, and environmental variables. And more speculative questions about such things as the origins of art, language, and consciousness can only be approached from a perspective of co-adaptation in which behavioural, cultural, and biological factors can potentially interact.

Anthropology as a discipline gains in significance precisely by providing a point of articulation between the long term and short term through its concern with both biological and socio-cultural processes. While that does not mean that every anthropologist shares the discipline-wide concerns he or she belongs to a polythetic set which overlaps with other sets across the discipline. This eventually results in the sharing of ideas and all too rarely the development of collaborative projects. Research by social anthropologists on marriage patterns, as Smith (Chapter 1) and Dunbar (Chapter 3) argue, shows how certain individuals gain selective advantage through preferential reproduction and how certain systems of marriage might facilitate the material well-being of a population. Such research shows the ways in which human social practices could potentially articulate with processes of biological adaptation. However, there are very few examples that show selection operating to the adaptive advantage of a population as a whole. Distributional studies demonstrate very close correlations between certain environmental variables and skin colour. But, as Harrison points out in Chapter 2, although it is possible to show the advantage that dark skin may confer in tropical environments, we are a long way from explaining the overall pattern of distribution. 'Sickle cell' anaemia, as Smith shows, is one of the few clear-cut cases where a mutation may have been selected for its adaptive advantage in reducing the impact of malaria on individuals and in turn on the population. Yet with reference to distributional studies in general Smith concludes that 'there is nothing in these analyses of gene frequencies and language which encourages a belief that adaptation is in general other than a second order effect in the determination of genetic variation' (p. 21).

In discussing genetic variability in human populations in Chapter 3 Dunbar stresses how a particular adaptive advantage may be only one factor among many that influence the selection of a particular gene. There are often a number of different ways in which the biological body can respond to a particular circumstance or satisfy a particular requirement. None of these may confer an absolute advantage over another. Furthermore a particular gene may be adaptive not because of any single effect but because of its contribution to the whole organism. Dunbar goes on to say that 'in biology an optimal solution to a particular problem need not always be a perfect solution because of the need to achieve a working compromise for the system as a whole.' Sickle cell anaemia again provides a classic example: the deleterious effect of the gene in its homozygous state (where the same characteristic is inherited from both parents), is more than compensated for by the protection conferred to heterozygote individuals against malaria. In the case of sickle cell anaemia the compromise is visible to the analyst and the adaptive significance of the gene can be demonstrated. However in most cases the compromises and adaptations are deeply embedded in evolutionary history and extremely difficult to disentangle.

What happens at the genetic level is thus particularly difficult to unravel; what happens at the phenotypic level is often clearer. This is nowhere more true than in the analysis of behaviour, where examples of adaptive behaviour are ubiquitous. Great progress has been made in the understanding of animal behaviour, especially social behaviour, in the last twenty years through the development of theories of kin selection, reciprocal altruism, and evolutionary strategy, and these are now increasingly being applied to human behaviour. It is through social behaviour that human beings have become so dominant a species. There are clearly many more complexities to human behaviour than to that of other animal groups but socio-biology has already provided a stimulating input to the debate. These issues are taken up in more detail by Robin Dunbar in Chapter 3.

The application of genetic or biological models to cultural and social forms has been a particular area of controversy, in particular given the difficulty of demonstrating specific instances of biological adaptation. The qualifications that Dunbar introduces – questions of co-adaptation, the existence of alternative responses, multi-determinacy, compromise – apply equally if not more so to the development of cultural forms and social behaviour. Dunbar (Chapter 3) and Morphy (Chapter 4) both draw attention to Dawkins's concept of *meme*, a unit of cultural information or an idea, proposed as a cultural equivalent to the gene. Human beings certainly develop bodies of cultural knowledge and social practices that enable them to respond to environmental circumstances by factors as variable as selective breeding, crop rotation, transhumance, and age-grade organization. Adaptation and adaptability seem to characterize much of human history. In some cases a particular form of life may appear to have an advantage over another, with populations and economies replacing one another over time. But it is open to debate whether the genetic analogy is helpful in understanding the mechanisms involved. The language of analogy leads inevitably to the danger of imposing a way of thinking about a topic that is not founded in the analysis of substantive data but rather rests selectively on apt illustrations. Dawkins's *memes* are so like genes that the biological could simply be substituted for the cultural and vice versa: 'Imitation, in the broad sense, is how memes can replicate. But just as not all genes that can replicate do so successfully, so some memes are more successful in the meme pool than others. This is the analogue of natural selection' (Dawkins 1989: 195).

The broad human capacity to acquire useful adaptive knowledge and to pass on information through the short cut of culture, rather than having continually to reinvent the wheel, is certainly a product of biological evolution over time. But the process of transmission itself and the particular properties of cultural knowledge are better conceived of in other than genetic terms. Just as Harrison argues that from a physiological perspective the human body has evolved a general capacity to respond to or adapt to changing climatic and environmental circumstances, so also have human beings evolved the cognitive capacities which allow them to utilize cultural resources to adapt to different environmental circumstances. The interaction between culture and genetics may have involved a degree of co-adaptation in particular cases and at particular points of time, but co-adaptation does not require that the specificities of cultural transmission have to be modelled as if they were genetic. The gene for lactase tolerance may have evolved in an environment in which adults selected milk as a food, but while the population does not have the option to reject the gene it could cease, for cultural reasons, to consume milk.

If culture is treated as genetic then it is difficult to avoid models in which ideas and artefacts apparently breed over time, producing increasing efficiency based on functional criteria. There is a danger that those cases where such factors can be demonstrated will be emphasized because they fit the model. In some cases functional efficiency is indeed the determining factor in the final form of an artefact, but in most cases artefact form is determined by multiple factors which in addition to function include style, status, aesthetics, and so on. The main objection to a genetic model of culture itself, as opposed to the human capacity for culture, is that cultural forms are transmitted extra-somatically and are not embodied except in a metaphorical sense. While areas of culture may be patterned, while technological change may have a trajectory that has a degree of directionality to it, the ideas which constitute culture are passed on by processes that are external to the body of knowledge itself, by the human beings who use the knowledge for particular purposes many of which are not straightforwardly functional. Likewise in the case of social behaviour such as marriage patterns, the articulation of the social with the biological can be taken into account, without reducing the one to the other.

Studies of human adaptation and adaptability require an understanding of both biological and socio-cultural processes and the development of analytic frameworks capable of including both kinds of data within the same model. The negative response of some social and cultural anthropologists to biological models and in particular to neo-Darwinian models of culture is partly a reaction to the intellectual climate of the nineteenth century in which simple evolutionary models were applied to explain all variations in human biology and behaviour often on the basis of environmental determinism. The model of cultural process developed by Pitt Rivers in the 1870s shows some similarities to that developed by Dawkins a hundred years on, though based on a much less sophisticated evolutionary theory. 'Modifications in the form of [artefacts] result from the succession of ideas or other causes affecting particular minds. They obtain acceptance by the survival of the fittest [idea]' (Pitt Rivers 1906: 26).

Evolutionist anthropology was rejected at the end of the nineteenth century both because so little that could be observed could be explained in terms of its principles and because it was associated with a unilinear theory of racial progress. Subsequent developments in knowledge about human biology have long since overturned the racial models which some nineteenth-century culture theorists adhered to. It became evident that very little of human genetic variation was explicable in terms of adaptation, and that racial divisions are more a social artefact than a biological one. Indeed, as Smith suggests, what is most significant about the human genetic inheritance, for the purposes of understanding adaptation, is what we share in common not what differentiates us from one another.

Rejection of nineteenth-century evolutionism should not, however, lead to a denial of the importance of adaptation as a process. That is to throw away the baby with the bath water. In the long term, adaptation has been an important process in the development of people as they are, and the study of human adaptability is an important part of both biological and cultural anthropology. Long-term processes of adaptation involving cultural and social behaviour have resulted in human beings who are extremely flexible organisms capable of adapting both culturally and physiologically to different environments and situations. Answers to the evolutionary questions of how we came to be the creatures we are require long-term collaborative research in which social, cultural, and behavioural factors must inform biological models and vice versa.

In conclusion, it is time to turn again to issues of complexity and scale. In the short term what characterizes human populations – genetically, physiologically, behaviourally and culturally – is their adaptability. Almost by definition that adaptability, and the fact that change and flux are integral to human existence at all levels, means that it is difficult to define determinate causes of what we observe.

However, to reject the concept of adaptation because human beings are so adaptable would be absurd. One of the challenges is to show what the constraints on human adaptability are, how adaptation occurs in particular contexts, and how adaptability at one level of existence relates to adaptability at others. Living in extreme environments – at altitude, in deserts, in polar regions – requires physiological, behavioural and cultural adaptations. How the body adapts physiologically may influence the cultural and behavioural adaptations that occur and vice versa. Considering adaptation in one area without reference to the others will produce only a partial picture.

We cannot expect to uncover causal relations often, except at the most general level, for example: warmth must be maintained in order to survive a polar winter! However, it is possible over the long term to see trajectories and structures in the development of regional systems, technologies, marriage alliances, and cultural forms which have a relative coherence and which represent the way in which a particular population reproduces itself over time – in short, its mode of adaptation. Technology but equally, as Howard Morphy shows in Chapter 4, the cultural values that are part of a way of life constrain the possibilities that can occur, and influence how people can behave. Most of the variation that occurs as a result of human adaptation is of no demonstrable evolutionary significance in that it has no direct genetic correlates. Nevertheless, over the longer term the expression of human adaptability in context, against the background of human genetic variation, is precisely the arena in which the mechanisms of evolution have operated, by advantaging an individual who is better adapted to the environment than another.

References

Dawkins, R. (1989). *The selfish gene.* Oxford University Press.

Pitt Rivers, A.H.L.F. (1906). *The evolution of culture and other essays* (ed. J.L. Myers). Oxford University Press.

Reynolds, V. and Tanner, R.E.S. (1995). *The social ecology of religion.* Oxford University Press.

ACKNOWLEDGEMENTS

We acknowledge gratefully a grant from the Parkes Foundation which enabled the proceedings of the Biosocial Society Workshop to be published.

CONTENTS

CONTRIBUTORS

Robin I.M. Dunbar
Department of Anthropology, University College London, Gower Street, London WC1E 6BT.

G. Ainsworth Harrison
Institute of Biological Anthropology, University of Oxford, 58 Banbury Road, Oxford OX2 6QS.

Howard Morphy
Pitt Rivers Museum, University of Oxford, South Parks Road, Oxford OX1 3PP.

Malcolm T. Smith
Department of Anthropology, University of Durham, 43 Old Elvet, Durham DH1 3HN.

1
GENETIC ADAPTATION

Malcolm T. Smith

Introduction

One of the achievements of Darwin's theory of evolution by natural selection was to provide an alternative explanation of adaptation, a mechanistic means by which the appearance of purposive functional design had been achieved without a designer. Our initial apprehension of human adaptation seems to me still to be a reaction to the obvious or supposed functional design aspects of our form and habit. Comparison with our closest ancestors and relatives plays a part here in influencing our judgement. Our large brain, our upright stance and bipedal gait, the functional competence of hand and eye, surely these are adaptations which have led to our success as the most numerous, widespread, socially complex, and cleverest primate of all.

I have used 'adaptation' above both to signify a state of adaptedness (as in 'human adaptation', which does not refer to any particular trait, but to a general condition) and to indicate some particular character of an organism which enables it to function in the constraints of its environment, and thus to achieve that state of adaptedness. This seems to be the intuitive usage of Darwin himself when, for example, he speaks of the giant land crabs of Keeling Island in the Indian ocean, which lived on a diet of coconut, as 'as curious a piece of adaptation and instinct as I ever heard of' (Desmond and Moore 1991, p. 182).

Usage is not always quite so simple, nor uniform, but I hope its complexities will not be problematic here; indeed it may be that a brief account of the terms 'natural selection' and 'fitness' in relation to 'adaptation' will help to explain the purpose of this essay, and the means by which it is to be achieved. Dobzhansky (1972 and elsewhere) considered 'adaptation' to be the process by which a state of adaptedness was achieved, whilst a structure or function contributing to adaptedness (an adaptive trait) might also in practice be called 'an adaptation'. In Dobzhansky's all-embracing view adaptation might occur through genetic change or physiological plasticity, though only the former could be considered an evolutionary process (rather than an evolutionary outcome). In a major review, Endler (1986) retained Dobzhansky's usages, and elaborated considerably. In the evolutionary sense, then, complex adaptation is achieved by natural selection. By 'natural selection' I mean the process that occurs as a consequence of variants in the population having hereditary traits which endow them with differential survival and fertility in the prevailing environment. It is the cumulative effect of this process which best explains the origin of complex functional design through adaptation. Contrary possibilities have been rigorously scrutinized by Dawkins (1986) and demonstrated to be wanting. Natural selection is often spoken of as a force which acts on genetic variation, but the above description makes it clear that this view is misleading: natural selection is no more a force acting on genes than 'overtaking' is a force acting on cars travelling at different speeds. Endler (1986) discusses these issues entertainingly and at some length.

If we focus on the genic level, however, both the 'complexity' and the 'design' which characterize phenotypic adaptation may be missing. At this level an adaptive trait (one conferring improved survival and reproductive success) may be simple, in terms of DNA perhaps no more that a single codon different from the alternative allele. Whilst such a trait arises by mutation, may increase in frequency, and be maintained by natural selection, there is also a role for random effects. Chance may increase the frequency of neutral or even deleterious alleles, and it can eliminate even advantageous mutants (Li 1976, after Fisher 1930). Although the average effects of random genetic drift are predictable, the outcome of individual cases is not. Nevertheless, we should expect in general that an advantageous mutant would be selected for, in other words, it would have a higher 'fitness' than a relatively disadvantageous allele at the same locus. The meaning of the term 'fitness' varies between authors, but here I use it in the sense of 'relative fitness' or 'Darwinian fitness', the extent that one trait relative to another becomes represented in the next generation through the differential survival and reproductive success conferred as a consequence of that trait.

At the genetic level, then, we are inclined to deem a trait adaptive not through recognition of its design to fulfil a function, but as residual demographic effect of its performance of that function, namely its Darwinian fitness. This effect, essentially that of changing the scale of resolution from the phenotypic to the genetic, has been paralleled by the work of Hamilton (1964), Williams (1966), and Dawkins (1976), among others. Their work has brought about a shift of emphasis in evolutionary biology at all levels by adding to, or even replacing, the criterion of functional design by the yardstick of reproductive success. The extension of evolutionary thinking to the analysis of human behaviour has widened still further the concepts of adaptation. Caro and Borgerhoff Mulder (1987), for example, describe as adaptive both traits which are shown to confer reproductive success (with or without a hereditary basis or genealogical transmission), and traits which look as though they would confer reproductive success on the basis of their design features. The latter argument certainly plays its part in the analysis of morphological and genetic traits, but it is an approach which is often inconclusive, as evidenced by debates on the reproductive function of all manner of traits, from colour blindness (Post 1962; Malhotra 1978) to the clitoris (Gould 1991).

In addressing human genetic adaptation in this essay I shall focus on the following issues. To what extent can it be demonstrated that the human genetic endowment is adaptive, both in the sense that is it designed by natural selection to function within the constraints of the environment, and in the sense that the relative fitnesses of genetic attributes can be ascertained?

There are at once two limitations to be mentioned, which follow from the way microevolutionary studies have developed, and which I am unable to overcome. The first is a constraint on the level at which adaptive explanations have been commonly sought. In the neo-Darwinian view there is a presumed genetic underpinning of adaptive morphological characters; we are adapted at a genetic as well as a morphological level. It does not automatically follow, though, that every 'gene', however construed, should demonstrably confer some advantage, only that no gene should confer a disadvantage such that the general functionality of the organism is compromised. Indeed, some authorities have argued that the adaptive genetic level is the genome rather than the locus (Mayr 1970; Lewontin 1974). In Mayr's view the harmonious interaction of loci is achieved as the outcome of generations of selection, producing an adapted phenotype which may lose fitness either through some unconditional deleterious event, such as the segregation of a lethal recessive, or through a loss of coadaptation of the gene pool effected by mutation or migration. In spite of this view that the effect of a gene is dependent not only on the organism's environment but is also conditional on its genetic background, the theoretical approaches and empirical genetic studies considered below have invariably examined individual alleles and single, or at best two, loci, on the basis of 'other things being equal'.

The second limitation is contextual, and again contingent on the historical development of the subject. That is to say, the systematic revelation of the extent of human genetic polymorphism occurred during a prevailingly selectionist climate when the views of Mayr (1963) and Ford (1964), for example, were very influential, and the subsequent literature reflects this bias and, to a lesser extent, reaction against it. Whilst acknowledging the positive stimulus of neo-Darwinian theory in anthropological interpretation, we must beware of letting adaptation be accepted as the 'common sense' default assumption. Something of this aura of 'But surely we're adapted!' still pervades and perhaps motivates the invocation of adaptive explanations. Conventional wisdom changes, however, and during this century opinions as to the maintenance of variation and the taxonomic value of traits have diverged greatly and altered swiftly. Successively, morphological 'racial characteristics' and then blood groups have been hailed as taxonomic markers on the express grounds that they had no adaptive function, and thus would not have been subverted by natural selection from yielding true indications of phylogenetic relationship. Boyd (1973), for example, argued for a different recent common origin of the morphological similarity of African Pygmies on the one hand, and the Andaman Negrito and Papuan 'Pygmies' on the other, on the grounds that their divergent ABO, MN, and Rhesus blood group frequencies belied their common morphology. Not only has the adaptive value of categories of variation been debated, it has often been stated that morphological characteristics (through being polygenic) are more reliable indicators of phylogeny than single gene traits, on the grounds that they are less susceptible to drift. This assertion, aside from begging the enormous question of why random change is not an adequate or even superior indicator of evolutionary divergence than selective change, has been shown to be a conjecture without foundation (Jorde 1985).

To elaborate further the point made at the beginning of the previous paragraph, to some extent the attempt to explain human genetic adaptation in adaptive terms should be seen as a battle in a larger war: the conflict over whether genetic polymorphism is for the most part accounted for by random fluctuations in the frequencies of functionally equivalent (or 'neutral') alleles, or whether polymorphisms exist because genetic variation within populations is actively maintained by natural selection. The shadow of that debate falls over almost all the case studies to be discussed below, and indeed is in part responsible for the fact that the quest to demonstrate natural selection and adaptation has been overwhelmingly the search for an explanation of genetic variation within the human species. This must be set against the view that selection on the species as a whole (stabilizing selection), observed when chromosomal anomalies, mutations, and variants beyond the viable range are eliminated, is both uncontroversial and essentially trivial to demonstrate. This does not gainsay its profound importance, nor diminish the sheer scale of its effect. I shall return to this issue in the concluding discussion. Meanwhile, these constraints mean that our search is confined mainly to the morphological or the single locus level, with variation within the species providing most of the empirical evidence.

The detection of selection

To what extent can it be demonstrated that the human genetic endowment is adaptive, both in the sense that is it designed by natural selection to function within the constraints of the environment, and in the sense that the relative fitnesses of genetic attributes can be ascertained?

A number of authors have reviewed methodology for the detection of selection and fitness differentials. The two fullest treatments are included in Lewontin's book *The genetic basis of evolutionary change* (Lewontin 1974), and Endler's monograph *Natural selection in the wild* (Endler 1986). Similar taxonomies of approaches are given by Morton *et al.* (1966) and Neel (1975). The classification employed here has been largely drawn from a reading of these authors, whose views overlap but by no means coincide. The headings used in this essay are based both on the different kinds of evidence available and also on the more pragmatic criterion of the amount of evidence available. Where a topic has been analysed exhaustively, such as the relationship of genetic variation to malarial distribution, I have been inclined to treat it separately, even though it might conceptually be considered within a broader theme. The following headings, then, set out the arrangement of material in this essay.

Ideally, one would like to be able to show not only that a trait performs a particular biological function, but also that through performance of that function the fitness of bearers of the trait is consistently increased or diminished. In practice such fitness differentials are very hard to demonstrate conclusively, and surrogates or components of fitness are often used instead. These might be, for example, survival to reproductive age, longevity, completed family size, distortion of segregation ratios, and so on, and all such will here be subsumed under the blanket term of 'Demographic measure'.

Whilst fitness differences may be hard to demonstrate, they might be predicted from the functional biology of variant traits. This demonstration of functional differences is most convincing when it occurs *in vivo*, but in many situations the only evidence we have is on the basis of *in-vitro* activity. These issues are addressed principally under the heading of 'Predictions from trait biology'.

A further category of information which can be used for inference of selection can broadly be called distributional. Within this a number of separate empirical correlates of genetic variation can be identified, yielding evidence which is more or less convincing: I have included under 'Empirical study of distributions' a consideration of global genetic distributions independent of any particular environmental correlation, and various correlates of genetic variation such as the physical environment, linguistic variation, and disease, including deleterious recessives. Malaria is considered as a subject in its own right, partly because there is such a wealth of detail available for this one disease and its relationship with genetic variation, but also because of the symbolic significance of malaria as a model selective agent. Because the explanation of recessive genetic disease frequencies has come to have a special significance for the inference of balancing selection, such diseases are examined under a separate heading, 'Inference from recessive disease frequency'. A quite different kind of evidence is exhibited in a consideration of 'Deviations from formal models', reviewing studies in which models predicting the distribution of genetic variation under defined circumstances are tested against actual genetic data, and the underlying assumptions evaluated.

Finally, under the catch-all heading of 'Miscellaneous features', I have devoted a little space to interspecific comparisons, changes through time, and incompatibility selection. None of these yields very much compelling information, but they hint at possible selective inference.

If not from the outset, it will be clear as we proceed that these categories are not in fact exclusive, and that certain traits or situations crop up under more than one heading. Where there are, indeed, independent lines of evidence pointing to the same conclusion, this can only increase our confidence that selective action is revealed.

Demographic measures

A number of approaches have been used to obtain evidence of selection from demographic evidence. I shall consider here deviation from Mendelian segregation ratios, evidence from differential fertility, from gene frequency differences between age groups, and from the Crow's index of the opportunity for selection. As will become clear, in general these methods are dogged by the need to examine very large data sets in order to infer selection at levels which may be realistic in evolutionary terms (Neel 1989). The unfortunate consequences are that studies of practicable size usually fail to detect selection, and that they also leave appreciable levels of selection undetected. For instance, Reed (1975) has shown that in searching for differential fertility or differential rates of spontaneous abortion in relation to parental genotype, rates of selection of up to 9 per cent and 28 per cent respectively could have gone undetected in a survey of 4576 couples.

Deviation from Mendelian segregation ratios

Mendel's laws predict the frequency of progeny genotypes among the offspring of particular matings, and one of the causes of distortion of these predicted segregation ratios is natural selection. Other possible causes include technical errors in identifying genotypes, non-disjunction, and meiotic drive. Among selective explanations, heterozygote advantage, incompatibility selection, and pre-zygotic selection are all candidate choices. There is a fairly extensive though rather dated literature of attempts to illustrate these, though the difficulties of organizing such a study are exemplified by Bodmer's (1968) calculation of the very large numbers required to detect small departures from Mendelian offspring ratios, with about 300 000 children needing to be observed to be reasonably sure of detecting a 1 per cent departure from a 1:1 ratio due to selection.

Several papers have addressed the issue of whether the MN blood group polymorphism is maintained by selection. Perhaps the most powerful is by Morton and Chung (1959), who found that among a large series of published studies from Europe and America there was a highly significant excess of MN children among the offspring of MN intercross and backcross matings, apparently limited to the children of MN mothers. Although technical errors could not be ruled out, the authors gave reasons for their belief in the remoteness of this possibility. However, subsequent studies have not upheld this claim, and the very authoritative voices of Race and Sanger (1975) concluded that the apparent excess was indeed due to grouping errors and that 'the subject is best forgotten' (Race and Sanger 1975, p. 120).

Distortion of segregation ratios has been sought in the ABO blood group system, but more influential still have been reported deficiencies of offspring numbers owing to incompatibility selection. Thus, Morton *et al.* (1966), in a study of more than a thousand families from north-eastern Brazil, reported a highly significant effect of blood group A incompatibility, together with an excess of heterozygotes in compatible test-crosses. Incidentally, this same study failed to find reproductive effects or heterozygote excess in any other of the 29 alleles under investigation. Cavalli-Sforza and Bodmer (1971) discussed the sampling requirements for comparison of the progeny of blood group genotype AO male × OO female and AO female × OO male matings in order to detect ABO incompatibility selection (on the assumption that there would be a shortfall of AO progeny born to OO mothers). They calculated that to demonstrate a 1 per cent disadvantage of blood group A 320 000 offspring were needed; to demonstrate a 10 per cent disadvantage 3200 would suffice. These figures are of an order beyond the scope of most studies and ambitions, but in reality the situation is still more discouraging. The theoretical calculations are of immense value, but they cannot of themselves take into account the kind of socio-economic confounding variables which inevitably will intrude in a large survey. Even surveys of limited extent struggle to match samples by socio-economic criteria, and in studies where sampling tends towards comprehensive enumeration the problem is inescapable.

An additional problem which has to be faced in applying demographic models at a family level is that of confidence of paternity. The incidence of non-paternity, when a woman's husband or partner is not the biological father of her baby, is a subject about which a certain amount of anecdotal information exists (Diamond 1991 provides just such an anecdotal account*), and which has recently been reviewed by Macintyre and Sooman (1991). From a genetic point of view non-paternity due to a woman's intercourse with another man or through artificial insemination is no different from step- or adoptive fatherhood, but it is much less likely to be declared in surveys. Macintyre and Sooman (1991) quote published rates in England which range from 30 per cent, based on Philipp's study of antibody formation in more than two hundred families in southeast England to 5 per cent, observed by J.H. Edwards in a study of ABO and Rhesus blood group incompatibility, fertility, and sex ratio in 2596 babies and their families. The authors state that medical students are usually taught that the rate is between 10–15 per cent, and that frequent assertions of a rate of 10 per cent appear to be based on 'hearsay, anecdote, or unpublished or unevaluable findings'. The importance of all this for the study of selection and adaptation by demographic methods is simply that in a field of study where required sample sizes are inhibitingly large, the existence of even 5 per cent uncertainty of paternity erodes still further the efficacy of the procedures involved.

Differential fertility

McCluer (1978) used computer simulation to assess the probability of showing differential fertility in genetic studies. Her approach was to generate from her computer program FERTSIM cohorts of 200 women with different levels of fecundability or different amounts of early fetal loss, and to see whether analysis of variance could detect differences between them. Various measures of fertility—including the total number of live births, the interval between birth and marriage, the birth interval between successive live births, and the woman's age at the birth of her

*'That study's findings have never been revealed until now. I recently learned those facts from the distinguished medical scientist who ran the study. (Since he does not wish to be identified in this connection, I shall refer to him as Dr. X.) .... The study's research plan was straightforward: go to the obstetrics ward of a highly respectable US hospital; collect blood samples from 1000 newborn babies and their mothers and fathers; identify the blood groups in all the samples; and then use standard genetic reasoning to deduce the inheritance patterns. To Dr. X's shock, the blood groups revealed nearly ten per cent of those babies to be the fruits of adultery!'

last liveborn child—were employed. The total number of live births was found to be the most sensitive indicator of differences in fecundability, though for the cohort size examined, even a 20 per cent reduction in fecundability was detectable only about 50 per cent of the time. This is not an encouraging prospect for the detection of realistically small differentials.

Age-group differences

Hiorns (1972) and Hiorns and Harrison (1970) have addressed the issue of inference of selection from variation in gene frequencies between age-groups. Hiorns envisaged the situation where it was not possible repeatedly to visit a study population to perform a longitudinal study, but instead a single sampling event was all that could be obtained to provide information. In these circumstances the sample could be stratified into age groups and inference made on the basis of genetic differences between the resulting strata. Two distinct situations for the comparison of age-groups were considered by Hiorns and Harrison (1970). The first represented a single generation, in which the younger age group had not yet undergone selection but the older group had; the second represented two generations, in which the younger group was the as yet unselected cohort of offspring produced by the post-selection older group. Although the two schemes produced considerable differences both in the size of selection coefficients that could be detected and in the range of recessive gene frequencies over which detection was possible, this variation paled beside the general inference that the sample sizes required were large. This inference held whether it was directional selection against a recessive or balancing selection maintaining a polymorphism that was under consideration. In the latter case, Hiorns (1972) explained that 'even for quite substantial selection coefficients of ten or twenty per cent, the sample sizes required would have to be of the order of several thousands in order to establish the existence of selection using a five per cent significance level for the test concerned'. The maximum sample size considered by Hiorns and Harrison was 8000 in each age-group, and the authors acknowledged that for many non-Western populations the census size is well below this. In respect of directional selection, they concluded that at levels of selection below 10 per cent against a recessive gene of frequency around 0.1 or less, selection was in practice undetectable by an age-group approach, whichever of their research designs was employed.

Opportunity for natural selection

Crow's Index of Opportunity for Natural Selection (Crow 1958, 1989) was devised as a measure of the total amount of selection which would take place if all pre-reproductive mortality and the variance in reproductive performance were genetically determined. This explicit assumption (which is acknowledged in the tag 'opportunity') does not, of course, obtain, and for this reason among others the measure has been criticized (Morton 1970). The index is computed as the relative increment in fitness in one generation. It is clear that the greater the proportion of liveborn children dying before reproductive age the greater the amount of selection that may occur. It is less intuitively obvious that because variance in fertility contributes greatly to the index, the greatest possible amounts of selection may be computed for post-demographic transition populations where infant and child mortality is low, but where there is a low mean and high variance in progeny size. The real difficulty in translating this index into inference about selection lies, as has been stated, in the fact that the heritability of pre-reproductive mortality and of fertility is uncertain, and that whilst some premature deaths and some cases of childlessness are without doubt genetic in origin and selective in effect, others, almost certainly the majority, are clearly not. A good review of the subject with exemplary case studies is given by Spuhler (1976), and more recent references can be found in a study of the Utah Mormons by Jorde and Durbize (1986).

In conclusion of this section it is as well to reiterate what many have acknowledged, that owing to the sampling criteria discussed above the demographic approach to selection will always be severely constrained. There was an early and thorough discussion of the issues in Neel and Schull (1968), and the imaginative exploration of ways to study natural selection combined with a realistic appreciation of the demands of demographic methods has been a continuing theme of Neel's work (Neel 1975, 1989). I shall leave the last words on this subject with Cavalli-Sforza and Bodmer (1971, p. 315), who likewise conceded the scale of the task imposed by demographic methods: 'It is a paradox of evolutionary theory that the small differences that may be so effective in molding the future of a species are so difficult to measure directly.'

Predictions from trait biology

Aspects of human morphology readily suggest that they are the result of design to fulfil function. Organs of apparent perfection such as the human eye, for example, are the classic ground over which arguments of purposive creative design versus adaptation through natural selection have long been disputed (Darwin 1859; Dawkins 1986). What has not been in dispute is that the eye *appears* to be designed to fulfil its functions.

It is tempting to extrapolate from the most convincing examples of functional design and to assume uncritically that many other aspects of human morphology are the result of adaptation. Especially engaging in this regard are traits which show marked geographical variation in their distribution. The task here has been to show that the variants are each adapted to their own environmental circumstances. The factors which bear on the matter as selective agents are the perceived features of the environment. I shall begin by discussing some of the classic morphological traits, before turning to ecogenetic and pharmacogenetic markers.

Body form

A very substantial analysis of geographical distribution of phenotypic variation in body shape in relation to possible selective and functional significance has been performed by Roberts (1973*a*). The analytical technique involved was to sample from the literature data on population values for various morphological measures, and to plot these against mean annual temperature, using correlation and linear regression to interpret the plots. Although there was an enormous amount of variation about the regression lines, some trends did clearly emerge. Linearity of phenotype, as measured by relative span, was correlated with mean annual temperature. Relative sitting height, which measures the length of the legs compared to the body, and is thus inversely related to linearity, decreased with mean temperature. Measures of absolute size such as weight and chest girth are related to the volume:surface ratio, and these were also negatively correlated with mean temperature. Roberts concluded that these trends are consistent with the general patterns of distribution of animal size and shape in relation to the environment that are encapsulated in Allen's and Bergman's Rules. Notwithstanding acknowledged phenotypic plasticity, it can be inferred that on a global scale these aspects of morphology are in great measure genetically controlled, and that the present world-wide distribution among indigenous peoples is the result of adaptation to maintain body temperature in different regimes of heat stress.

Speculation has abounded about other morphological aspects of the phenotype, and several have been claimed as adaptive. Thus, hair colour and texture, head and face shape including nasal breadth and prominence, eye colour, epicanthic eye fold, and steatopygia have all been with some plausibility claimed to be adapted to the environment. The problem with these explanations is that they are for the most part retrospective justifications not based on experimental evidence of survival value or reproductive success. Roberts (1973*a*) quoted some data on the cooling of parts of the face which lent some support to the view that the Mongoloid face—with flat nose, and fat padding over eyes and cheeks—might well protect from frostbite and warm inspired air. There is also evidence that the shape of the nose is important in humidifying inspired air.

Skin colour

Skin colour is a multifactorial trait with considerable phenotypic plasticity for which perhaps no more than four loci are responsible (Harrison 1988, Cavalli-Sforza and Bodmer 1971, p. 527). Darwin (1874) suggested that universal interest in skin colour as a measure of beauty, together with the fact of pigmentary variation between populations (but not between sexes) might point towards pigmentation as an outcome of sexual selection rather than adaptation to environment. This claim has been voiced again recently by Diamond (1991) on the grounds that the complexity of suggested adaptive explanations is bewildering, and that correlations of pigmentation with incident sunlight are too imperfect to carry conviction. However, a thorough review by Robins (1991) makes the case that at least some environmental stresses are likely to exert a considerable selective effect. Firstly, we must acknowledge that the global distribution of skin colour among indigenous peoples is correlated with latitude in a way that cannot be explained by a hypothesis of sexual selection alone. Secondly, testable mechanisms have been put forward to explain both the advantage of dark skin in the tropics and of light skin in temperate regions. The two most developed of these concern the causal role of sunlight in skin cancer and in vitamin D synthesis. The following accounts rely heavily on Robins (1991), who should also be consulted for references.

Skin cancer is more common in fair-skinned Caucasians, especially of Celtic descent, than among Africans, Amerindians, and Asians. Onset of skin cancer is not related to age, but to the cumulative amount of exposure to UV light, and the occurrence of skin cancer significantly affects survival. Studies of albino negroes showed that cancers or premalignant lesions occurred in all subjects by the age of 20 years, and that 6 per cent of Nigerian albinos were in the age range 31–60 years compared to 20 per cent of the non-albinos. Additionally, melanin screens the body from the sunlight, and the pain and discomfort of sunburn and its after-effects might seriously impair performance among hunter–gatherers, thus exerting a selective effect.

Vitamin D is an important promoter of calcium absorption from the gut and of normal calcification of bone, and a deficiency of vitamin D in infants and children causes the defects in bone structure called rickets. Rickets was widespread in the polluted cities of England and Europe during the industrial revolution, and in the last thirty years has been encountered in Britain mostly among Asian communities. The principal source of vitamin D is by synthesis in the skin in the presence of UV light, but it may also be obtained from components of the diet, notably fish oils and, in the modern world, fortified food. From the point of view of vitamin D synthesis, light skins, including albino phenotypes, are not disadvantaged in the tropics, since there is no occurrence of over-production of vitamin D and resultant toxicity.

A vitamin D-based evolutionary hypothesis to account for the distribution of light skin colour away from the tropics may be summarized as follows. Pale skin is adaptive in temperate climates with low incident sunlight because it permits adequate synthesis of Vitamin D, without which there would be high rates of rickets, and consequent problems not only with hunting and gathering but also with reproduction, owing to malformations of the pelvis. Robins finds this hypothesis inadequate on several grounds. Firstly, experiments have shown that enough vitamin D is synthesized even as a result of brief exposure to sunlight and even at northern latitudes. Secondly, rickets seems to be a disease of urban, overcrowded conditions, where subjects have been deprived of natural sunlight through pollution or cultural practice. Thirdly, archaeological evidence of rickets is uncommon in prehistoric Europe and absent in pre-Columbian America, when the hypothesis suggests that selection should have been acting against darker skins.

With the vitamin D hypothesis thus disposed of, there is no shortage of other possible suggestions to explain the adaptive role of skin colour variation, for example the alleged resistance of depigmented skins to cold stress or the claimed protection afforded by dark skin to the destruction of folic acid by UV light. We begin, however, to share Diamond's frustration that the survival value of a feature commonly thought to be adaptive is so far from adequately demonstrated.

Ecogenetics and pharmacogenetics

The significance of ecogenetic and pharmacogenetic variation is twofold. In respect of some characters we can observe polymorphic variation which, through ecological correlation of known functional variants, is by inference adaptive. Much less compelling are other cases where the demonstration of adaptation or selection cannot directly be made, but we can appreciate the possibility of selection occurring, through showing that different genotypes respond differently to pharmacological stimuli or have detectable differences in activity.

Persistent adult intestinal lactase (PL)

Lactose, the sugar present in milk, is digested by the enzyme called lactase. The discovery that in many populations of the world adults do not have this enzyme followed efforts from the United States and Europe to use milk powder as part of food supplementation programmes to help tropical populations where there were food shortages. Baker (1988) records that by the 1960s it was widely observed that such supplementation resulted in diarrhoea among the recipient population, and that this could be shown to be a specific reaction to ingesting milk.

Systematic testing showed that in many Asian, native American, and African populations only up to a few per cent of adults have lactase, whilst in European and some African populations almost all adults do so. The feature which makes PL of adaptive significance is its distribution. By and large PL occurs in populations with a tradition of milk use, the exceptions being some PL deficient populations which utilize cheese and yoghurt, which have no lactose (Kretchmer 1972).

As to the genetic control of PL, Cook (1978) reports a study which suggested an autosomal dominant mode of inheritance inferred from the children of European–Yoruba marriages, though it has also been suggested that continued exposure to milk after weaning may maintain the enzyme (Baker 1988). One factor which argues against a purely genetic determination of the trait is the existence of widely differing frequencies of PL in neighbouring populations (for example the Ibo with no PL and the Fulani with 80 per cent). It is unlikely that differences as profound as these could have accrued by natural selection during the limited period of isolation of related populations, especially given that the maximum time of domestication of sheep and cattle is of the order of 5000 years (Johnson *et al.* 1981).

Whilst the initial selective advantage to PL might be to exploit the milk from other mammals during a food shortage, there have also been a number of second order arguments put forward, such as the ability of milk to reduce the growth of intestinal bacteria, or to enhance the absorption of calcium in environments with low ultraviolet light and low dietary vitamin D. Cook (1978) argued on the basis of PL incidences and population history that the trait might first have evolved among desert nomads of the Arabian peninsula, where the ability to drink camels milk would give not only a nutritional advantage but also an increased intake of water and electrolytes.

Phenylthiocarbamide (PTC)

Phenylthiocarbamide is one of a number of related substances which some people taste as intensely bitter, whereas to others it is tasteless. Population studies on these reactions revealed one of the earliest known human polymorphisms, having two alleles at a single locus, with tasting ability a Mendelian dominant. Research into patients suffering from various thyroid diseases has revealed an inconsistent but striking pattern of association with this polymorphism and dietary sources of iodine. Mourant *et al.* (1976), who provide a useful short review, are of the opinion that the polymorphic frequencies might be maintained by balancing selection.

N-*Acetyl transferase*

N-Acetyl transferase is important for the acetylation of many drugs, and people exhibit one of two phenotypes, fast or slow. Epidemiological studies suggest that slow acetylation predisposes to arylamine-induced bladder cancer. Within the group of people classed as 'slow acetylators' predisposition is variable, with greater susceptibility reported as acetylation capacity diminishes (Cartwright *et al.* 1982).

α_1 *Antitrypsin*

This trait is referred to variously by the title above and as proteinase-alpha-1 inhibitor, abbreviated as PI or Pi. The principal phenotype of α_1 antitrypsin deficiency is chronic pulmonary disease or emphysema, through the destruction of pulmonary alveoli. There are more than 70 electrophoretically distinguishable alleles in the European population. The commonest allele is PI^M, with a frequency of 0.95. Most of the disease caused by α_1 antitrypsin deficiency is due to two alleles, PI^S and PI^Z, of which the latter in its homozygous form is by far the more dangerous, reducing α_1 antitrypsin activity to 10–15 per cent of normal.

Alcohol and aldehyde dehydrogenases

The enzymes responsible for the metabolism of alcohol vary between individuals and populations, and are responsible for differing degrees of tolerance to (and perhaps preference for) alcoholic drinks. The metabolic pathway involves alcohol dehydrogenase (ADH), which converts the alcohol to aldehyde, and aldehyde dehydrogenase (ALDH), which breaks down the aldehyde to carbon dioxide and water. (Small amounts of ethanol are excreted in urine, sweat, and expired air.) There are a number of ADH loci of which ADH_2 and ADH_3 are polymorphic. At the ADH_2 locus the proportion of persons with the 'abnormal' phenotype varies from 5 per cent in the English, 20 per cent in the Swiss, and of the order of 90 per cent among the Chinese and Japanese (Agarwal and Goedde 1991). Although *in-vitro* differences between the phenotypes have been demonstrated (Kalow 1991), the two forms are thought to be equivalent in function for the *in-vivo* metabolism of ethanol.

However, there do seem to be physiological and behavioural correlates of the variation in aldehyde dehydrogenase activity which have been reported by Agarwal and Goedde (1991). Deficiency of $ALDH_2$ activity has been shown at levels of between 8 and 53 per cent among Asian populations, at about 40 per cent in South American Indians, and 2–5 per cent in North American Indians. This enzyme deficiency is absent from populations of European or African origin.

$ALDH_2$ deficiency is correlated with alcohol sensitivity symptoms such as facial flushing, increased heart rate, and palpitations. Among Japanese subjects, flushers with $ALDH_2$ deficiency had higher blood levels of acetaldehyde than non-flushers, but did not differ in their blood ethanol levels (Harada *et al.* 1985). It also seems likely that $ALDH_2$ deficiency plays a role in alcohol use and abuse. For example, in a study of 101 Japanese there were 19 per cent regular drinkers among the 42 per cent who were $ALDH_2$ deficient, but 49 per cent regular drinkers among the 58 per cent who had the ALDH normal phenotype. Furthermore, significantly reduced levels of $ALDH_2$ deficiency have been observed among alcoholics compared to controls in a Japanese psychiatric hospital.

Cytochrome P-450 enzymes

The inducibility of aryl hydrocarbon hydroxylase (AHH) has been reported to increase the risk of lung cancer related to cigarette smoking. An early model of one locus with two alleles and three phenotypes (low, intermediate, and high inducibility, conferring a one-, 16-, and 36-fold increase in susceptibility) has failed to be confirmed (Calabrese 1984). Recent work suggests that at least two loci responsible for cytochrome P-450 enzymes are implicated in sensitivity to cigarette smoke. In the P-450 I family the P_1 enzyme is in essence what was earlier described as AHH (Calabrese 1991), whilst in the P-450 II family the defects in the enzyme debrisoquine 4-hydroxylase have also been associated with bronchogenic carcinoma in smokers. Debrisoquine metabolism deficiency, due to a number of different mutations, may be exhibited by 5–10 per cent of the population.

The studies of human morphological variation make a strong case for the existence of adaptive hereditary variation resulting from natural selection. They also suggest, however, that the appearance of functional design is insufficient grounds alone for an understanding of the determining processes. At the genetic level, allelic variations in function at a large proportion of polymorphic loci (Harris 1975) together with the striking differences in reaction to many pharmacological stimuli (Grandjean 1991) seem to me strongly suggestive of adaptive possibilities. With a trait such as persistent adult intestinal lactase, the issue is beyond dispute:

this clearly is a case of genetic adaptation, and a particularly good human example, too, with cultural behaviour intervening between the species and the environment.

Empirical study of distributions

There is more distributional evidence about blood group and enzyme polymorphisms, in particular the ABO blood groups, and about their associations with diseases, than about any other genetic traits; the enormous reviews by Mourant *et al.* (1976, 1978) bear incontrovertible witness. This emphasis will shift rapidly as more DNA-based population studies are carried out at the population level (see, for example, Wainscoat *et al.* 1986; Bowcock *et al.* 1991) but for the time being the distributional data on gene product polymorphisms is unsurpassed. A number of conjectures have been made on the basis of these 'conventional' gene frequency distributions as to the microevolutionary processes responsible. Brues invoked stabilizing selection to explain why the global distribution of ABO frequencies was so restricted, and founder effect has been suggested to account for depletion of alleles in the New World and Australia. The clinal variation in ABO frequencies across Europe and Asia has been interpreted as a response to selection by infectious diseases (Vogel 1975) and also, in parts of its distribution, as a result of migration (Haldane 1940; Sokal 1989). The overall pattern of gene frequencies in Europe is much more complicated, however, and not consistent with any single explanation; demic expansion, mosaic settlement of migrants, and latitudinal selection are all likely to have played a part (Sokal 1989). In the following paragraphs I shall consider aspects of geographical distribution, and the relationship of gene frequency variation to languages and to disease.

Spatial patterns

An interesting comparative approach to the distribution of blood groups, electrophoretic polymorphisms (serum proteins and enzymes), and HLA genes was followed by McClellan *et al.* (1984). As part of their study of genetic distances in Mormons, they computed genetic distances between Mormons, other religious groups in the USA (Amish, Hutterites, and Mennonites), and 13 European populations. Genetic distances were calculated from three separate sets of data, six red cell antigens (ABO, MNS, Rh, Kidd, Kell, and Duffy), nine electrophoretic systems (ACP, C3, HPA, ESD, GLO, GC, PGM1, BF, and GPT), and HLA-A and HLA-B loci. For each data set, the complicated resulting pattern of pairwise distances was plotted in two dimensions using as axes the first two dimensions of the genetic covariance matrix. The resulting plot can be read like a geographical map (Jorde 1985). In the present case the electrophoretic and HLA-based genetic distance maps for the European populations clearly resembled a geographical map of Europe, and suggest that isolation by distance (migration, gene flow, and drift) were the main determinants of genetic structure. This view was strengthened by the position of the religious communities, with the Mormons (having a large founding size and many subsequent immigrants) similar in gene frequencies to the European mainland, and the Amish, Hutterites, and Mennonites (characterized by small founding populations and a subsequent history of isolation) showing marked divergence from the European populations and from each other. By contrast, the genetic distance map based on red cell blood group genetic distances distorted the geographical distances, focusing most of the populations much more closely together, with just Iceland, Poland, and Finland as outliers from the main group. This difference between the plots, based on HLA and electrophoretic systems on the one hand and red cell blood groups on the other, may suggest that the blood groups in European populations have been under stabilizing selection, constraining them towards a European ecological norm. Free from such constraint, the other loci have differentiated in accordance with isolation by distance predictions.

The second point I want to make about spatial distribution of gene frequencies is that there is there is a convention in the explanation of human genetic variation that historical causes be adduced. This is quite different from the case in explaining variation in plants and animals, where a historical perspective is often impossible, and the convention of explanation is much more slanted towards selection. The point is well made in Lewontin (1974). The case study used here to illustrate this is a local one, but the general argument is, I believe, more widely applicable.

ABO blood group distribution in Britain has been thoroughly documented by Kopeć (1970), and is clearly clinal, with higher group O frequencies in the north and west, and with just occasional discontinuities in the distribution (Roberts 1953; Watkin 1960). This pattern of ABO blood groups is almost always explained in terms of differential regional settlement by invading populations possessed of different gene frequencies (Brown 1965; Roberts 1973*b*; Watkin 1986; and see Lewontin 1974, for a particularly succinct and dogmatic statement of the case). It is perhaps worth pointing out, though, that archaeologists have shifted from models of invasion towards regional continuity in the explanation of the distribution of artefacts, and a selective causation for the present ABO distribution in Britain might plausibly be contemplated. On the assumption that phenotypes involving B and A alleles are more resistant to infectious disease than O and that the prevalence of infectious disease is positively associated with population density we can adopt an extremely simple model of selection in which all phenotypes except blood group O are fully fit, and the O phenotype has a selective disadvantage proportional to the local density of population. To assess the possible effect of the Black Death on gene frequency variation in England Smith (1989) explored such a model using estimates of medieval population density in AD 1086 and in AD 1377 made by Russell (1948), and found that both were consistent with the observed present-day distribution of ABO blood groups. The significance of this is not that it proves selection to be responsible, nor that it falsifies the conventional explanation couched in terms of settlement migration, but rather that it shows the normal interpretation to be chosen rather than entailed. In the preceding example population density and infectious disease were chosen as plausible causal correlates of genetic variation which might stand as an alternative to prehistoric and historic settlement. These, however, are simply exemplary, for it has long been observed that there are many other systematic geographical correlates of geography in Britain, such as altitude, temperature, rainfall, and geology (Fox 1943; Blackman *et al.* 1963).

Correlation with language

Recent research comparing the linguistic and genetic affinities between populations on a global scale has shown a remarkably strong correlation between the two patterns of relationship. Cavalli-Sforza *et al.* (1988) (see also Cavalli-Sforza 1991) showed that a phylogenetic tree based on gene frequencies in 38 populations revealed the same major groupings as a classification of their languages. Although the linguistic taxonomy is not formally an evolutionary classification, the authors concluded that the same principal mechanism—the generation of divergence in isolation—was responsible for population affinities in both spheres, and suggested a parallel evolution of genes and languages. Subsequently, the evidence for this coevolution has been shown to be highly statistically significant (Cavalli-Sforza *et al.* 1992).

Moving down in scale from the world-wide to the continental, a number of studies have confirmed the importance of language as a major correlate of genetic frequencies. There is a substantial amount of recent research on the distribution of gene frequencies in Europe which indicates the importance of linguistic boundaries and colonizing settlement. For example, Bertranpetit and Cavalli-Sforza (1991) have analysed the distribution of blood groups, enzymes, and proteins in the Iberian Peninsula using a technique which produces what they call 'synthetic genetic landscapes' which are the result of a geographical mapping of the highest principal components of gene frequencies. These components are independent, and demonstrate independent patterns in rank order of weight. The first component, i.e. the single most important factor in the differentiation of the Iberian Peninsula, depicts the Basque versus non-Basque distinction, with the marked genetic differentiation of the Basques in the present day being attributed to the persistence of characteristics of the Mesolithic population which are probably due to genetic drift. This view of the Basques as an ancient relict population is also entirely consistent with linguistics and physical anthropology.

In an important series of papers, Sokal and his co-workers have undertaken to evaluate the importance of linguistic boundaries in explaining genetic variation. A number of methods have been employed, including 'wombling' (named after its inventor, Womble), which is designed to detect abrupt changes in the pattern of distribution of biological variables (Barbujani and Sokal 1990). By this technique 33 genetic boundaries were detected, of which 31 were associated with language. Other techniques include the comparison of gene frequency variance in quadrats containing a linguistic boundary with that in quadrats which are linguistically homogeneous, and the measurement of rates of change and of gene frequency difference across 29 language–family transitions (Sokal *et al.* 1988). This battery of techniques reveals European gene frequency variation to be much less smooth than would have been the case if isolation by distance were the sole influence. At the same time, the overwhelming correspondence of genetic change with linguistic rather than physical boundaries or ecological markers suggests that adaptive processes were relatively unimportant, with the major contribution to genetic boundaries being the immigration of populations of diverse origin.

At a more local level still, Barbujani *et al.* (1992) have shown that the distribution of genetic variation in Italy differs from predictions based on a model of isolation by distance. Using wombling, they demonstrated a number of lines of genetic discontinuity which coincide with linguistic and linguistic/geographical boundaries. Even at a scale of resolution of only tens of kilometres Bittles and Smith (1990) have shown in the Ards peninsula of Northern Ireland the existence of a gene frequency boundary influenced by a settlement history with strong linguistic and religious components. In this case, though, as perhaps in the case of genetic differentiation in Italy, the linguistic difference is just one cultural marker among several which maintain but did not initiate the genetic divergence.

In the Americas there has also been strong interest in the relationship of genetic variation and language. Salzano and Callegari-Jacques (1988) compared pairwise genetic distances among 41 South American groups belonging to 10 linguistic stocks, and showed that with the exception of the Tupi speakers there was always a lower mean genetic distance within linguistic stocks than between them. These results are consistent with several other studies in the New World, which have shown associations between linguistic and genetic variation among Eskimos (Crawford *et al.* 1981), North American Indians, and between and within South American tribal groupings. However, Salzano points out that several other studies do not show this pattern, and he reflects that linguistic relationship often points to a common environmental history, with the implication that genetic correlations with linguistics should not automatically be interpreted as non-adaptive consequences of isolation between groups.

On balance, then, the distributional evidence associating gene frequency differences with linguistic differences between populations is ubiquitous and clear. Its interpretation, however, is not quite so straightforward. On the one hand, the analytical studies of Sokal *et al.* (1988), Barbujani and Sokal (1991), and Barbujani *et al.* (1992) on the correlation of genetic change with language suggest that historical settlement is a more powerful determinant of genetic differentiation than isolation by distance or natural selection. On the other hand, the correlation with language does not exclude adaptive effects, either independently of the linguistic influence, as when clinal variation at a locus runs counter to a language boundary, or subsumed within it through the shared environmental history of a group speaking closely related languages. The most comprehensive surveys seem to suggest a multiplicity of causes, as remarked by Sokal (1989). Nevertheless, there is nothing in these analyses of gene frequencies and language which encourages a belief that adaptation is in general other than a second order effect in the determination of genetic variation.

Correlation with disease

There are two aspects of the relationship between genetic markers and disease which might be revealing of natural selection. First is the association within populations of particular genotypes with diseases, from which might be inferred differential susceptibility and subsequent selective elimination. I shall consider ABO and HLA associations with a number of infectious and organic diseases. Second is distributional evidence about the varying incidence and severity of disease between populations, with resistance based on postulated or observed gene frequency differences between the populations, and conjectured to have arisen through preadaptation, or through selection by similar environments in the past. Under this heading I shall consider population associations with leprosy and tuberculosis.

Some of the first reported associations between the alleles of polymorphic blood group systems are those concerning the ABO blood groups with duodenal ulcer (Aird *et al.* 1953) and with carcinoma of the stomach.

These associations have been very influential as icons against neutralism, although even ardent selectionists such as E.B. Ford acknowledged that conditions such as cancer of the stomach, which occurred after the period of child-bearing, could not exert selection to lower the frequency of the susceptible blood group (Ford 1975, p. 125). Cavalli-Sforza and Bodmer (1971, p. 303) addressed the question of natural selection in relation to such disease associations, and in the case of duodenal ulcer they calculated the fitnesses of O and non-O phenotypes. The incidence of duodenal ulcer in females of blood group O relative to women of other blood types is 1.4. In spite of this strong association between the O phenotype and duodenal ulcer, the reduction in fitness of O females was calculated to be no more than 6.4×10^{-6}. There are two main reasons why the association is not translated into a substantial effect on fitness. Firstly, the overall incidence of duodenal ulcer is low, and secondly, the disease tends to affect older, people, causing very few deaths among those of reproductive age. Manly (1985, p. 370) repeated Cavalli-Sforza and Bodmer's analysis, quoting a correction to their procedure offered by Nei and Yokoyama (1981), and showed the relative fitness of O compared to other phenotypes to be 0.999985.

It is clear from these analyses that the demonstrated associations, though of clinical significance, do not have important consequences for microevolution. Generally acknowledged to be of much greater potential are associations between genetic markers and infectious diseases, because in this case many people may be affected, they may be affected at an early age, and they may die as a result of their infection.

A large number of studies have sought associations between ABO phenotypes and infectious diseases, with smallpox, plague, syphilis, influenza, tuberculosis, leprosy, and infant diarrhoea among the conditions receiving attention. Initial reports were very stimulating, with Vogel *et al.* (1960), for example, hypothesizing that global ABO frequencies could be explained by susceptibility of blood group O individuals owing to the plague bacillus, *Pasteurella pestis*, possessing H antigen. Likewise, a claimed higher relative incidence of smallpox among people of blood group A and AB stimulated a great deal of research. Unfortunately, subsequent results have not presented a clear picture, and even enthusiastic proponents have more recently reviewed the situation as unproven (see, for example, Vogel 1975). In spite of the enormous amount of data that has been collated (Mourant *et al.* 1978), strong support has not emerged for this kind of simple adaptive role for blood groups, and most authorities treat the claims with caution (Bodmer and Cavalli-Sforza 1976, p. 340; Harrison 1988, p. 274). More realistic than this simple model of constant adaptive values attributed to individual alleles might be one of frequency dependent selection (Clarke 1975), where common alleles were the target for pathogenic specificity and rare alleles consequently advantageous. The dynamic of the latter situation would maintain polymorphism, whereas, in the absence of other effects, the former would erode genetic variation (Cavalli-Sforza and Bodmer 1971, p. 212).

The HLA complex includes a number of loci on the short arm of chromosome 6, and represents about 0.001 of the human genome (Thomson 1988). Three regions have been defined, of which Class I includes the HLA-A, -B, and -C loci, Class II the HLA-DP, -DQ, and -DR loci, and Class III contains complement and cytochrome genes. Genes at the Class I and II loci are highly polymorphic, and associations between HLA haplotypes and over 40 diseases have been established. In spite of the fact that some of theses associations are very strong, for example those with ankylosing spondylitis and with insulin dependent diabetes mellitus (IDDM), no direct evidence of these associations with viability or fertility has been recorded (Bodmer 1973). Only in the cases of C2 deficiency and congenital adrenal hyperplasia is the mechanism of disease association understood, and for other associations it is conjectural whether the HLA loci themselves are implicated through their function in immune defence, or whether unidentified closely linked loci are responsible (Ryder *et al.* 1981; Thomson 1988).

Other evidence that the HLA complex is implicated in selection comes from studies of genotype distributions (see below) and from the existence of linkage disequilibrium. Bodmer *et al.* (1988) argued that the extensive linkage disequilibrium between HLA-B and -C loci, which occurs in all major population groups, is probably the consequence simply of very close linkage and little consequent recombination, but that the strong disequilibrium in European populations between alleles A1 and B8, which are not so closely linked, is probably maintained by some form of selection.

Leprosy

Spickett (1962) argued on the basis of variation between populations in the incidence of leprosy, that one of the factors involved was genetic susceptibility. For example, he showed that in Brazil the incidence among native Brazilians was significantly lower than among Brazilians of foreign extraction. Likewise, an isolate of Bavarian descent in Colonia Tovar, Venezuela showed far higher levels of infection than those prevalent in the surrounding indigenous population. Within populations, too, there is a tendency for the disease to be limited to certain family lines. This constraint is the more noticeable in populations where the disease is rare, and has been noted in Europe, Iceland, and the United States. The data are ambiguous, however, being consistent with a contagion theory of causation as well as with a genetic hypothesis. From family studies

Spickett concluded that a single irregularly dominant gene, with penetrance of around 80 per cent, which had the effect of making the body susceptible to invasion by *M. leprae* could be invoked to explain the observed pattern of transmission in an extensive pedigree from New Brunswick.

Schistosomiasis

It has been suggested by Abel *et al.* (1991) that there are genetic differences in predisposition to high infection intensities of *Schistosoma mansoni* in Brazil. The inference of genetic effects was on the basis of segregation analysis on 20 Brazilian pedigrees, adjusting for other (than genetic) predisposing factors, such as sex, age, and contact with water. Woolhouse (1992) has commented that although the effects of age, sex, and exposure to the infective agent have been accounted for, the inference of a genetic component to susceptibility may not be warranted, since the residual family effects (interpreted as evidence of a major co-dominant gene for susceptibility, having a frequency of 0.20–0.25) might actually be mediated through some small scale differences in the ecology of families rather than by genotype.

Tuberculosis

Motulsky (1979), among others, has commented on the fact that Jewish mortality rates from tuberculosis have been significantly lower than in other populations, in spite of the fact that the level of infection appears to be similar among Jews and non-Jews. This has encouraged the hypothesis that among Jews there is a genetic basis to resistance. The existence of some genetic factors operating in resistance to tuberculosis has been indicated by twin studies, and this fact gives some plausibility to Motulsky's claim, though he concedes that factors other than genetics might well be responsible for interpopulation differences in susceptibility. According to Sawchuck and Herring (1984) the suggestion of a selective mechanism by which Jewish populations acquired resistance was first proposed by Fishberg in 1911, and has subsequently been reiterated in numerous publications cited by them. Fishberg's claim was that tuberculosis mortality in the ghettos of medieval Europe had eliminated all but the most resistant genotypes from the Jewish population.

Sawchuk and Herring's own analysis of tuberculosis in the Sephardic Jewish community of Gibraltar confirmed the reported mortality differential between Ashkenazi Jews and non-Jews, but through their detailed consideration of socio-economic variables these authors were able to point to a number of confounding influences. Not only did the Jewish families in Gibraltar enjoy higher material and social status, their occupations were such as not to aggravate tuberculosis. Thus, over half the non-Jewish labourers worked with coal, tobacco, or in stone masonry; none of the Jewish workers was so employed. Furthermore, there was a conscious concern for health care among the Sephardic Jews, including a charitable institution for the sick and aged, which might have lessened the impact of tuberculosis in the Jewish community.

If we concede that there might be some genetic differences in resistance to tuberculosis either within or between populations, there still remains the question of identifying a candidate gene or genes and a mechanism responsible for resistance. The area which has received most attention is that of overdominance, with advantage supposed to be conferred on the heterozygotes for various deleterious recessives present at beguilingly high frequencies in Jewish populations. Meindl (1987) has argued that the high frequencies of the cystic fibrosis gene in European populations might be the result of selection by tuberculosis, whilst the Tay Sachs Disease (TSD) gene has also been proposed as a candidate by Myrianthopolous and Aronson (1966). In addition to TSD, which has a frequency 0.17–0.4 births per thousand in Ashkenazi Jewish populations (Weatherall 1991), three other genetically quite distinct lysosomal storage disorders characterized by excess sphingolipids (important in cell membranes and in the central nervous system), Gaucher disease, Niemann-Pick, and mucolipidosis IV, have each been shown to reach high frequencies in particular Israeli communities (Zlotogora *et al.* 1988). The presence at unusually high frequency of four different yet related diseases in Jewish isolates populations might be explained on the shifting balance theory as a case of locally adapted optima in populations evolving in isolation (Hartl and Clark 1989).

Inference from recessive disease frequency

The search for candidate genes naturally leads to an examination of any genes present at unusually high frequency, and thus involves as a class deleterious recessive genes, whose frequencies would be otherwise predicted to be kept low by mutation–selection balance, but yet might be elevated through heterozygote advantage. Any high-frequency deleterious recessive, then, draws speculation about the adaptiveness of its heterozygote. Two more such recessives are considered below.

One would expect the frequency of deleterious and especially of lethal recessives to be low, with homozygotes eliminated by selection and largely maintained by segregation from matings between heterozygotes. It is, however, a matter of observation that the frequency of many recessive diseases varies between populations. For example, the frequency of Tay

Sachs disease is elevated in Ashkenazi Jewish populations, phenylketonuria (PKU) in Northern Irish, cystic fibrosis in European, and, of course, HbS in African populations and populations of African descent (see below). Genetic isolates are also characterized by high frequencies of otherwise uncommon alleles, often deleterious ones. Higher than expected frequencies in isolates, however, may be attributable to random effects as well as to balancing selection, among other factors; the problem lies in obtaining reasonably convincing evidence in favour of a particular cause.

Phenylketonuria

Phenylketonuria (PKU) is a recessive disease in which deficiency of the enzyme phenylalanine hydroxylase (PAH) results in an inability to metabolize the essential amino acid phenylalanine. The condition affects about 1 in 10 000 newborns in Europe and the United States, rising to about 1 in 400 in Ireland and 1 in 800 in Scotland and Scandinavia. Although PKU has been reported to occur at high frequencies in Gypsy populations (Williams and Harper 1977), 13 out of 14 copies of the gene in a Welsh gypsy kindred with exceptionally high rates of PKU could have been introduced by a non-Romany (Williams 1986). The gene has also been implied to have a Celtic origin (Saugstaad 1975, quoted in Williams 1986).

Recently Kidd (1987) has argued on the basis of RFLP studies in European populations and in Yemenite Jews that molecular heterogeneity at the locus supports a hypothesis of maintenance by balancing selection. On the basis of eight polymorphic RFLP sites, 12 unique RFLP haplotypes of normal PAH function have been defined (Mange and Mange 1990). Among these, Haplotypes 1 to 4 account for more than 90 per cent of the chromosomes in Danish kindreds. Haplotype 3, whilst accounting for 3 per cent of normal variation, is the origin of 38 per cent of the PKU genes in Denmark. Likewise, 20 per cent of the PKU genes derive from Haplotype 2. Kidd (1987) argues that these facts argue against multiple mutation as a cause of elevated PKU levels and diminish the probability that drift is responsible, since at least two mutations have achieved high frequency. This also rules out hitch-hiking by selection acting on a nearby unidentified locus. The residual explanation is balancing selection through heterozygote advantage, though no mechanism is suggested.

Cystic fibrosis

Cystic fibrosis is the most frequent lethal recessive disease in Caucasians, with a frequency of 0.4–0.5 per thousand births (Weatherall 1991), and about 0.01 per thousand births in African and Oriental populations. Attempts to explain its high frequency in Europe have adduced a number of independent causes. These include genetic drift, heterozygote advantage, reproductive compensation, high mutation rate, and multiple loci, all of which have been considered by Jorde and Lathrop (1988). In respect of heterozygote advantage, both survival and fertility differentials in favour of the heterozygote have been proposed. As regards survival, Meindl (1987) developed in depth the hypothesis that the selective agent associated with CF might be pulmonary tuberculosis. He argued that the epidemic which began in England in the sixteenth century and later spread to Europe and North America, intensifying with the urbanization of the industrial revolution in the second half of the eighteenth century, exerted a sustained selective effect for centuries. The advantage accruing to heterozygous carriers of the CF gene was argued to lie in their increased competence in synthesis of extracellular mucopolysaccharides by lung stroma fibroblasts, a feature that would be beneficial at the onset of the excessive fluid accumulation that characterizes tuberculosis infection (Meindl 1987).

In respect of fertility differences between heterozygotes and normals, an excess of children among the grandparents of children with cystic fibrosis has been reported (Danks *et al.* 1965; Knudson *et al.* 1967; Conneally *et al.* 1973). Although such an analysis removes reproductive compensation effects it leaves the possibility of ascertainment bias through fertility effects of the carrier grandparents. However, Jorde and Lathrop (1988) concluded on the basis of their detailed analysis of the Utah Geneaological Database that when ascertainment bias was excluded there was no evidence of fertility differences between heterozygous carriers and people without the gene. Likewise there was no difference in offspring birth interval between carriers and controls. The discussion of likely causes by these authors well illustrates that when the question of heterozygote advantage is raised, a number of rival candidate selective agents may readily be forthcoming. In addition to Meindl's advocacy of CF, discussed above, typhus, influenza, malaria, bubonic plague, and venereal syphilis have all been mooted (see Jorde and Lathrop for references).

Recently, molecular methods have enabled the discrimination of distinct CF alleles, and the population distribution of these has created a new focus of discussion. The following account, then, deals with the proportion of the most common CF allele among all CF alleles, rather than the distribution of cystic fibrosis itself. The most frequent CF allele in Caucasian populations is the DF508 mutation. De Braekeleer and Daigneault (1992) reviewed detailed evidence for its distribution in Europe, and showed a consistency between areas with persistent Celtic settlement and the highest proportion of DF508 among all CF genes. They postulated a single Middle Eastern origin for DF508, and both selective and founder effect to explain its high but variable frequency in a number of Caucasian populations. It is not so easy to explain on this hypothesis the high proportion of DF508 in the Basque people, where 87 per cent of all CF genes among Basques of unmixed ancestry are DF508 (Casals *et al.* 1992), unless one argues for initial gene flow from the non-Basque newcomers followed by genetic drift at high levels. Casals *et al.* favoured a hypothesis of Palaeolithic presence of DF508 in Europe, with dilution of its frequency with other CF alleles brought in during the Neolithic. Thus in the present day, the Basques would represent an ancient gene frequency distribution, with other European populations predictably lower in their DF508 frequency as a result of Neolithic settlement. However, the observed frequencies elsewhere in Europe (87 per cent in Denmark, 70–80 per cent in the populations of Britain, France, Belgium, western Germany, and the Netherlands, 50 per cent in Spain, 45–55 per cent in Italy) are not distributed in such a way that this could credibly be the only process involved.

For a recessive disease gene to be suspected of bestowing advantage in the heterozygote, it should not only be present at a frequency greater than the predicted mutation–selection balance in the absence of selection, but should also attain such frequency in a number of separate populations. By contrast, a higher than expected frequency of a deleterious allele in a single isolate does not strongly suggest selection, since, depending on the structure of the isolate, the outcome might much more simply have been achieved through a combination of founder effect, inbreeding, and drift. Indeed, high levels of some particular deleterious alleles are very characteristic of isolates as a whole. Thus in the North American religious isolate the Old Order Amish not only do there occur recessive diseases, such as cartilage-hair hypoplasia, which are extremely rare outside the group, but there are also examples of autosomal or X-linked recessive disorders which are frequent in one deme but not others of the distinct Amish populations: Ellis-van Creveld syndrome in Lancaster County, pyruvate kinase deficiency haemolytic anaemia in Mifflin County, and autosomal recessive limb girdle muscular dystrophy in Adams and Allen Counties (McKusick 1979). Similarly, French-Canadian and Finnish populations manifest a very unusual distribution of recessive disorders in comparison with their neighbours (Jorde *et al.* 1992; Veillette *et al.* 1992), and where detailed molecular and demographic studies have been done it seems that founder effect is responsible. Deleterious dominant conditions also reach unusually high frequencies in isolates. Examples include porphyria variegata in people of Dutch descent in South Africa (see Mange and Mange 1990 for references), myotonic dystrophy in the Sauenay-Lac-Saint-Jean region of Quebec Province, Canada (Veillette *et al.* 1992),

Machado-Joseph disease in the Azores and populations of Azorean–Portuguese origin (Boutté 1987), and von Willebrand disease in Sottunga, Åland Islands (O'Brien *et al.* 1988).

The presence of the same deleterious allele in a series of populations encourages us to think selection may be (or may have been) at work because the probability of the same allele increasing by chance in each one of them is vanishingly small. Thus, the high frequency of the recessive condition albinism among Hopi, Cuna, Jemez, and Zuni Indians (Woolf and Dukepoo 1973) encouraged explanations in terms of balancing selection with unknown agent, as well as a hypothesis of 'social selection'. The gene might be promulgated both because of the esteem in which albinos were held as a talisman, and because of the increased mating opportunity albino men might enjoy by not going to the fields to work with the men and boys but by staying in the villages with the women and performing weaving and other tasks (Woolf and Dukepoo 1973).

In the cases which have attracted most attention as indicators of an adaptive role for heterozygotes, namely Tay Sachs Disease and Cystic Fibrosis, such diversity of populations with high frequencies is indeed encountered. TSD, for example is elevated in many Ashkenazi populations, and Neel (1979) has argued that the case for a drift explanation is hard to accept, involving as it must high frequencies of the TSD allele in a particular pre-diaspora Jewish population of Palestine which then by chance went on to found the diverse and isolated Ashkenazi populations of Europe. A drift explanation involving the amplification of TSD frequencies after the diaspora, i.e. independently among the small, isolated, and inbred Jewish populations of Europe, is still less credible.

Malaria

The coincidence of a number of abnormal haemoglobins and other red cell anomalies with the global distribution of malaria constitutes the suggestive geographical evidence that malaria is an environmental agent maintaining these abnormalities at appreciable polymorphic frequencies by natural selection. The genetic systems involved include the abnormal haemoglobins S, C, and E, the thalassaemias, hereditary ovalocytosis, the enzyme deficiency Glucose-6-phosphate dehydrogenase, and the red blood cell shape anomaly, hereditary ovalocytosis. Conversely, in areas where malaria is not present, haemoglobin A is virtually monomorphic. This distributional pattern has been supplemented by demographic, physiological, and experimental evidence to produce a compelling argument for natural selection and adaptation.

HbS

At its simplest, the situation under discussion is one where there are two alleles (HbA and HbS) at a locus. The normal genotype is AA, the homozygote for the other allele (SS) confers severe anaemia resulting in premature death. The terms 'sickle cell disease' and 'sickle cell anaemia' refer to the characteristic sickle shape assumed by the red blood cells. The heterozygote AS condition (whose bearers are often described as having 'sickle cell trait') confers no ill effects in general, but may cause an adverse reaction under unusual conditions such as reduced oxygen pressure, or in response to certain drugs when a sickling crisis may be induced (Calabrese 1991). In the absence of some intervening agent of selection, we should expect the sickle cell gene to be present at very low frequencies maintained by the balance between recurrent mutation and the elimination of SS homozygotes by selection. What requires explanation then, is the elevated frequency of heterozygotes and SS homozygotes in malarial areas.

Livingstone (1983) records that the earliest suggestion of the malaria hypothesis is generally ascribed to Haldane (1949), though Durham (1983) and Serjeant (1992) cite observations made by Beet (1946) on the reduced frequency of malarial parasites in blood films from individuals with sickle cell trait. During the 1950s evidence of the close geographical association of the distribution of sickle cell trait and malaria had been published, and Allison (1954) showed that volunteers with sickle cell trait inoculated with *P. falciparum* parasites developed malaria less often and with less severe consequences than those without the trait. Other researchers found this latter observation not always to be reproduced, perhaps owing to the confounding of the effects of genetic resistance with acquired immunity. Livingstone (1983) alludes to a number of studies on parasite rates and densities which, taken together, provided only equivocal evidence of an advantage accruing to heterozygotes in this respect. Both Livingstone (1983) and Serjeant (1992) cite deaths from complications such as cerebral malaria to be less frequent among bearers of sickle cell trait, among whom the course of the disease is generally clinically uncomplicated. Durham (1983) concludes that with few exceptions, field studies report lower incidences of severe and fatal parasitism among AS genotypes compared to AA.

Among older children and adults in malarial areas there is an excess of heterozygotes and a pronounced deficiency of SS homozygotes, very few of whom survived to reproductive age. It has also been suggested that AS heterozygotes are more fertile than homozygotes (Livingstone 1983), either through a female advantage due to differential infection of the placenta, or a male advantage due to the effects of fever on spermatogenesis. However, Serjeant (1992) finds no evidence that this view has been convincingly demonstrated.

As far as the mechanism of resistance among AS heterozygotes is concerned, it seems that during the early stages of infection sickling is enhanced by the presence of the falciparum parasites. These sickled cells may be preferentially destroyed, but it also seems likely that the parasites within sickled red blood cells may be killed owing to low cellular concentrations of potassium ions (Nagel and Roth 1989).

The relationship between gene frequencies and rates of malarial prevalence has been investigated in detail by Durham (1983), using maximum monthly rainfall as a surrogate variable for the amount of malaria. Accurate gene frequency data for abnormal haemoglobins had already been collated by Livingstone from 157 populations. Durham showed by polynomial regression analysis that among Kwa, West Atlantic, and Mande language groups 78 per cent, 39 per cent, and 24 per cent respectively of the variance in HbS levels could be explained by rainfall (and thus by inference malaria). Exceptions to the general pattern of variation were encountered among genetic isolates, which had unexpectedly low HbS frequencies, and some other language groups, in which there was no consistent relationship between rainfall and HbS. Finally, there was a suggestion across all levels of rainfall up to 500 mm that higher frequencies of HbS occurred among the Kwa than among the West Atlantic speakers. Among possible explanatory causes was the possibility that the yams grown by the Kwa speakers reduced to some extent the sickling propensity of red cells, whereas the rice-based diet of the West Atlantic group had no such effect.

The malarial hypothesis grew up at a time in the development of neo-Darwinian theory when the prevailing mood was triumphally adaptationist and when the expectation of finding balanced polymorphism was high. As influential thinkers continued to stress the adaptiveness of variation itself, the scale and intensity of selection implied by the malarial hypothesis seemed to be the emerging vindication of this view. As such, the malarial hypothesis has, I believe, come to symbolize a model for the interpretation of other genetic polymorphisms despite the fact that as time goes by malaria is revealed more and more as unusual in the extreme.

Other abnormal haemoglobins

Haemoglobin C is widespread in Central West Africa, and its distribution has been taken to imply a selective advantage for the gene. Whilst the red cells of CC homozygotes are resistant to the growth of *P. falciparum*, and parasitized CC cells are resistant to lysis by comparison with parasitized

AA cells, there is apparently no difference between AA homozygotes and AC heterozygotes in their ability to resist infection by *P. falciparum in vitro*. Nagel and Roth report no resolution of this difficulty. They do suggest, however, that selective pressure to increase the HbC frequency might be through the double heterozygotes for HbS and HbC.

Haemoglobin E occurs at high frequencies in Southeast Asia. It is most common in northern Thailand and northern Cambodia, and decreases from there in all directions. Flatz (1967; reported in Lie-Injo 1976) noted that its distribution followed that of the Austroasiatic (Mon-Khmer) language group. It has been shown that parasitized EE and AE red blood cells are more readily phagocytized than AA.

Thalassaemias

The thalassaemias are a group of enormously variable defects in globin chain production, within which two major categories can be identified. Normally, equal amounts of α-globin and β-globin chains are produced, and they combine to produce the tetrameric haemoglobin molecules. In α-thalassaemia there is a relative deficiency of α-globin chains, and though the resulting excess β-globins do form homotetramers (called HbH) these do not function satisfactorily as oxygen carriers, resulting in reduced size and number of red cells, with consequent anaemia. In β-thalassaemia the β-globin chains are deficient, the α-globin homotetramers are insoluble and precipitate within the red blood cells, and anaemia again results, though in both thalassaemias there is a broad range of clinical effects. At a molecular level the conditions are very diverse, with α-thalassaemia (for which unequal crossing over between the two functional α-globin genes on chromosome 16 is a frequent cause) showing less heterogeneity that β-thalassaemia, in which more than 60 different mutations have been observed (Gehlerter and Collins 1990). For thalassaemia as a whole a total of more than 100 mutations have been recognized (Weatherall 1991). The thalassaemias have a global distribution covering the Mediterranean shores, Africa, the Middle East, India, and southeast Asia, closely matching world-wide pattern of malaria. Weatherall (1991) quotes a WHO estimate of 181.3×10^6 heterozygotes world-wide, contributing with HbS and other haemoglobinopathies to a total of 242×10^6. Thus the thalassaemias constitute the single largest category of haemoglobin defect.

The adaptive significance of thalassaemia as a protection against malaria is suggested not only by the coincidence of their distribution on a world scale, but also by local studies which have shown detailed spatial and altitudinal correlations. Such studies have been conducted in Italy and Sicily (Siniscalco *et al.* cited in Cavalli-Sforza and Bodmer 1971) and more recently, using molecular techniques, by Flint *et al.* (1986) in Melanesia. As regards a mechanism by which thalassaemia genes confer resistance, it appears that thalassaemic red blood cells retard parasite growth. The mechanism by which this is achieved is unclear, with *in-vitro* studies reviewed by Nagel and Roth (1989) including no fewer than seven rival possibilities.

G6PD deficiency

Glucose-6-phosphate dehydrogenase (G6PD) is an enzyme specified by a gene on the X chromosome. Although there are a large number of rare variants, two normal alleles predominate; these are designated B, which has 100 per cent activity and is common throughout the world, and A, which has 88 per cent activity and is confined to Black Africans and people of African descent (Mange and Mange 1990). Deficiency of this enzyme among indigenous peoples occurs in the eastern Mediterranean, sub-Saharan Africa, and parts of Asia. Deficiency does not cause chronic ill health in homozygous females and hemizygous males, but severe anaemia can result as a reaction to some drugs. Sensitivity to primaquine, for example, resulting in acute haemolysis, is caused by a variant of the A allele designated A^-, which has only 15 per cent of normal activity. The Mediterranean variant B^- (also called G6PD Mediterranean) has only 3–4 per cent of normal activity, and in some individuals can cause a severe acute haemolytic crisis as a result of eating fava beans, a condition called favism which was first recorded in the fifth century BC by the Greek historian Herodotus. Persons with favism may suffer haemolytic reactions to a number of other oxidizing substances, and also as a result of infections (for example typhoid fever, influenza, and viral hepatitis).

Malarial protection is afforded principally to heterozygous females, with the African genotype B/A^- being particularly favoured (Nagel and Roth 1989). *In-vitro* culture has clearly shown that G6PD deficiency inhibits parasite growth, but the mechanism remains unclear, though it may be that oxidative stress is particularly severe on parasites in G6PD-deficient red cells (Nagel and Roth 1989). Recent work at the molecular level has shown that the A^- variants are much more heterogeneous than had previously been thought, and Buetler (1989) raises the question as to why all the deficiency variants of African origin are mutants of the A allele, which currently has a frequency of only 20 per cent.

Hereditary ovalocytosis

Elliptocytosis or hereditary ovalocytosis is a dominant phenotype characterized by oval-shaped red cells, which are rather short-lived. The two conditions are distinguished by Nagel and Roth on the shape of the cells, though most other authors treat them as synonymous. It is determined by either of two loci, one linked to the Duffy locus (chromosome 1p) and the other linked to the Rh locus (chromosome 1q). In each case the elliptocytosis locus has been found to be close to the gene for an important red cell membrane protein. The phenotypes are indistinguishable except through linkage relations.

There is a high frequency of elliptocytosis among Aboriginal populations in Malaysia (Temuan, Jakun, and Senoi) whilst the frequency among Malays is very low (Fix and Lie-Injo 1975). Very high frequencies have been observed among the Toradja around Lake Lindu in central Sulawesi, Indonesia, where 'slightly less than half' the population had the condition (Lie-Injo 1976), and in samples from Kar Kar Island, Gogol Valley, and Gulf Province in Papua New Guinea 12 per cent of the population was affected. It is clear that in PNG the distribution of elliptocytosis is congruent with that of endemic malaria (Serjeantson *et al.* 1977) and in some populations reaches levels of 30 per cent (Amato and Booth 1977, reported in Motulksy, 1989) and Nagel and Roth (1989). In order to substantiate the implication of a protective role against malaria for this trait, Serjeantson *et al.* (1977) measured the parasite rate among people with and without elliptocytosis, and found that for all *Plasmodium* species combined (*P. falciparum*, *P. vivax*, and *P. malariae*) there was a slightly higher rate of infection (26.6 per cent) than among those with normal blood cells (24.1 per cent). Considering the effects on each species separately, the rate of parasite infection among ovalocytic children (age 2–14) compared to children with normal red cells was 1.05 for *P. falciparum*, 0.90 for *P. vivax*, and 0.50 for *P. malariae*.

Duffy null

Throughout most of the world the Duffy red cell blood group polymorphism is controlled by two co-dominant alleles (Fy^a and Fy^b) at a locus on chromosome 1. These alleles are responsible for the expression of Fy^a and Fy^b antigens on the red cell surface. Among African and United States black populations, however, and among other ethnic groups which may be admixed with them (Arabs, Oriental Jews, Druzes) there are substantial frequencies of the phenotype Duffy (−) (Duffy-negative), in which neither antigen is expressed. Mourant *et al.* (1976) showed that among most sub-saharan African populations the frequency of the allele (Fy) responsible for Duffy (−) is between 90 and 100 per cent. It is also the case that significant proportions of African and American blacks are resistant to *Plasmodium vivax* malaria.

In-vitro experiments revealed that Duffy (−) erythrocytes resisted invasion by the merozoites of *Plasmodium knowlesi*, a species closely related to *P. vivax*. *P. vivax* does not propagate in culture, and so Miller *et al.* (1976) used informed and consenting volunteers on whom *P. vivax*-infected mosquitoes were fed. The six men with Fy^a or Fy^b antigens contracted an erythrocyte infection 11 to 15 days after bites by infected mosquitoes. The five men with Duffy (−) were completely resistant to infection. The distributional and experimental evidence combine to suggest very strongly that populations with high levels of Duffy (−) have been subject to natural selection. Although *P. vivax* infection in otherwise healthy people rarely causes death, protection from infection is likely to be advantageous in childhood or at any time under nutritional stress. There is evidence that it is the Fy^a and Fy^b antigens on the erythrocyte surface that are involved in the invasion process, and may be the receptors for parasite attachment to the red cell.

HLA

Hill *et al.* (1992) reported that recent studies have indicated HLA associations with severe *P. falciparum* malaria. Three out of six healthy adults from an area of the Gambia with hyperendemic malaria, who had the HLA-B53 antigen, showed cytotoxic T lymphocyte (CTL) response to a particular peptide from liver-stage-specific antigen-1 (LSA-1), but not to sporozoite antigens. The effect was reduced in individuals with HLA-B53 from a less malarious part of the Gambia, and non-existent in two HLA-B53 residents of Oxford without exposure to malaria. It was suggested that the resistance mechanism is acquired during natural exposure to malaria and operates by action during the liver-stage of the parasites' life cycle. The effect is HLA-B53 restricted, and apparently peptide-specific, with CTL recognizing an epitope of LSA-1 which sequencing studies showed to be almost completely conserved in the Gambian parasite population. CTL activity to this epitope does not protect against other strains of *P. falciparum*, whose diversity in this respect may be maintained by frequency-dependent selection.

Deviations from formal models

This section gives an account of several models which predict gene and genotype distributions under certain assumed conditions, and which can be used to evaluate empirical data sets. I include for consideration the Hardy–Weinberg equilibrium, models based on Wright's *F*-statistics, and a simple model of admixture between populations.

Hardy–Weinberg equilibrium

The Hardy–Weinberg equilibrium is basic to the discussion of the behaviour of genes and genotypes in populations (rather than in families) and is considered at some length in all texts that deal with population genetics, for example Cavalli-Sforza and Bodmer (1971), Bodmer and Cavalli-Sforza (1976), Harrison (1988), Hartl and Clark (1989), Mange and Mange (1990), and Wallace (1981). The Hardy–Weinberg equilibrium describes the distribution of genes and genotypes under conditions when essentially no evolution is occurring: these conditions include among their stipulations that there be no mutation, no migration, no selection, no sampling effects, and no deviation from random mating. Although in practice most populations are in Hardy–Weinberg equilibrium, deviations from equilibrium may be due to a number of causes besides technical error, including inbreeding in its broadest sense, gene flow, selection, and random genetic drift. Deviation from Hardy–Weinberg equilibrium cannot of itself indicate that selection is responsible, only that selection is one of several candidate causes of the observed deviation (Endler 1986). Additionally, the equilibrium is very robust (Hartl and Clark 1989), so only deviations of considerable magnitude can be statistically detected. This decreases the chance of detecting selection in comparison with effects of a larger order, such as latent population heterogeneity (Cavalli-Sforza and Bodmer 1971). Lewontin (1974, p. 242) had no doubt that the appeal of the technique was that it required only a knowledge of different genotype frequencies among adults, but concluded that it was of limited application, yielding significant deviations only when there were very large viability differences between genotypes.

Methods based on F *statistics*

The Lewontin–Krakauer and Ewens–Watterson tests are members of a family of analyses of the distribution of heterozygosity and homozygosity in subdivided populations which are derived from Wright's *F*-statistics (Wright 1943; Neel 1975). Morton (1975), however, was already of the opinion that studies of the distribution of heterozygosity would be inconclusive owing to deviations in the real world from the assumed conditions of the models, and that the most that could be inferred was that mean systematic pressure per locus must be small. He concluded the article by saying that 'at the species level all evidence is consistent with the hypothesis that most polymorphisms are due to weak selection, whose effects are poorly distinguishable from neutral mutation' (Morton 1975, p. 152). Nevertheless, there have been a number of applications of these methods, and the technique continues to be employed.

Lewontin (1974) argued that if the variation in allelic frequencies among populations were due entirely to the breeding structure of the species, including the history of population size fluctuations and migration, then all loci should show the same degree of inter-population variation, as measured by the inbreeding coefficient, *F*. Measurement of *F* locus by locus should then result in a homogeneous set of *F* values with a variance no greater than that due to sampling error among the observations. If, however, only some loci were neutral, with the variation of others among populations determined by stabilizing selection (which would decrease total variance) or disruptive selection (which would increase total variance owing to local adaptation) or both, then the resulting set of *F* values would be expected to show a different variance. Lewontin and Krakauer (1973) calculated the theoretical variance of *F* values under conditions in which all variation was due to shared population history in order to test observed distributions of *F* against this null hypothesis. Bodmer has used this test to examine variation at the HLA-A, ABO, MNSs, Rhesus, P, Duffy, Kell, Haptoglobin, and Immunoglobin loci. The *F* values were very heterogeneous, with high variability between races and low variability within being interpreted as evidence of selection. Lewontin (1974) likewise used data on human blood groups and serum proteins in applying this method. He calculated that the observed data yielded a variance only a tenth the size of the theoretical expectation. He inferred that drift and migration were not sufficient to explain observed variation and that selection must be implicated. It was not clear from the data alone, however, whether the high *F* values for systems such as Rhesus, Gm, and Duffy were the result of diversifying selection, with the lower values representing population structural effects, or whether the low *F* values indicate selection towards a global equilibrium, with the high *F* values resulting from random differentiation. On external evidence, Lewontin chose the latter interpretation, in recognition of the deviant gene frequencies in small isolate groups such as Eskimos, Australian Aborigines, Papuans, and American Indians. However, Cavalli-Sforza and Bodmer (1971) pointed out that as well as correlations with population origin there are Gm correlations with climate; Cavalli-Sforza has also remarked (in the discussion at the end of Motulsky 1979) that the high frequency differences in Gm between populations indicate that the system must have been exposed to very rapid selection. Whatever the interpretation of this point, the method offers a criterion of selection based upon the distribution of *F* levels. It is, however, a method that has been much criticized, very succinctly by Endler (1986), for example, who says that there are serious problems in estimating the expected variance of *F* and in constructing a valid statistical test.

The Ewens–Watterson test is based on the infinite alleles model of genetic variation: at any locus there are an infinite variety of alleles possible, and the current distribution is purely the result of chance, each allele being assumed to function as effectively as any other. Ewens showed that if sample size and the number of distinct alleles are known, then the distribution of sample variation at equilibrium can be predicted in summary form according to an F value based on the squared allele frequencies (Hartl and Clark 1989). This measure is often referred to as 'homozygosity', but is homozygosity estimated from the assumption of Hardy–Weinberg equilibrium for each individual population. The F statistic measures the range of homozygosity observed across the sampled populations. The observed sample F can then be compared to an expected value which is usually obtained by simulation rather than analysis. Manly (1985) provides a FORTRAN program for this, based on an algorithm by Stewart (1977) (quoted in Manly 1985). F values higher or lower than expectation may be caused by selection.

The method works best on loci with large numbers of alleles, and has been applied to human data by Hendrick and Thomson (1983), who tested HLA-A and HLA-B locus data from 22 populations. Homozygosity at both loci was much less than that predicted on the basis of neutral alleles. The authors consider in some detail the factors which might be responsible for increased or decreased homozygosity. Agents responsible for the former include hitch-hiking, lack of equilibrium (perhaps due to selection in the recent past), and directional selection (including such selection as might be caused by disease associations). Causes of reduced homozygosity are stratification of the samples, such as might be caused by geographical sub-division of population, the presence of unidentified alleles, the existence of a population bottleneck in the past, or some kind of balancing selection, such as heterozygote advantage or frequency-dependent selection. On the basis of their consideration of these, Hendrick and Thomson concluded that some form of balancing selection is most consistent with the HLA variation observed.

A further application of the Ewens–Watterson test is provided by Hartl and Clark (1989), who quoted Clark's 1987 study of allele frequency distributions at the HRAS-1 locus in humans, identified by Southern blotting with the pLM0.8 probe and TaqI digests. In this study the F-test has not been performed, but rather the whole distribution of observed and expected allele frequencies are compared. The distributions are rather similar, but there is a clear excess of the most common allele; that is, the commonest allele is much more common than would have been expected by chance alone. Hartl and Clark (1989) claimed such a pattern to be consistent with stabilizing selection acting to eliminate rare deleterious alleles continually entering the population by mutation.

It would also be consistent with historical effects such as selection or drift, in which many alleles had been lost during some episode during the past.

Averill (1980) criticized both the Lewontin and Krakauer and the Ewens–Watterson procedures on the grounds that genetic drift can result in exactly the same genetic distributions that are potentially attributable to selection. The evidence in support of this contention was obtained from computer simulations in which only drift effects were input, but whose results were interpreted by the formal tests as indicative of selection. Averill concluded that the problem lies in the use of average drift effects, which predict a loss of genetic variation, rather than the realization of particular random outcomes.

Recently, some of the doubts expressed as to the adequacy of data have been met by Bowcock *et al.* (1991) who have tested Melanesian, African Pygmy, Chinese, and European populations for 100 DNA polymorphisms. The variation in *F* among these five populations was compared with predictions made under conditions of constant population size, with drift as the sole agent of genetic change. Deviations from expectation could not automatically be explained as due to selection, since a model of extreme fluctuations in population size might bring about the same effects. However, it seemed more likely that stabilizing or disruptive selection was affecting a maximum of one third of the genes under examination.

A further measure of average heterozygosity within species has been used by Nei (1988) to test whether the predictions of the infinite-alleles model of neutral mutation held in a sample of 30 species. In all species except one the observed heterozygosity was below that expected on the neutral hypothesis. Nei took this to indicate firstly that balancing selection could not be an explanation because deviation from expected heterozygosity was in the wrong direction (and the single exception to the direction of deviation was *E. coli* which, being haploid could not show overdominance). Secondly, he postulated that explanatory mechanisms might include either a Pleistocene bottleneck of population sizes due to glaciation, which might have reduced allelic diversity in a large number of species, or diversity-reducing selection, such as would occur if most mutant alleles were somewhat deleterious rather than neutral, or if selection was of fluctuating intensity. Disarmingly, Nei's strongest evidence of positive selection is that it *must* occur or there could not be morphological or physiological adaptation. He therefore envisaged one of two situations to reconcile the apparent conflict. Possibly most mutations were not strictly neutral and their cumulative effects over many loci resulted in adaptive change. Alternatively, most non-deleterious mutations were neutral, and there were a very few genes having major effect which were responsible for adaptive change, a view which Nei preferred.

Inference from admixture estimates

The estimation of admixture in hybrid populations, principally in the New World, has proved a popular application of elementary genetic theory of migration. The method requires that gene frequencies be known for two or more ancestral populations, and from this knowledge the proportion of each ancestral population in the hybrid one can be estimated. In order for estimates to be accurate it is desirable that the frequencies of an allele be as different as possible in the ancestral populations. Natural selection may be inferred from this method if one locus indicates a level of admixture apparently discrepant with the rate predicted by others. The odd one out may have been subject to selection. A good example is given by the paper by Gonzales *et al.* (1976) estimating admixture in a sample of 'negroes', 'whites', and 'mulattoes' in the population of Cuba. The category names appear in quotes because the analysis showed all three groups to be admixed! These authors found that among the mulattoes allele frequencies at three other polymorphic loci gave an estimate of 53 per cent Caucasian admixture, whereas sickle cell haemoglobin frequency suggested 79 per cent Caucasian genes in the population. The discrepancy was taken to indicate that in the non-malarious environment of the New World, HbS has been selected against, and thus is unrepresentative of the actual African admixture. This case seems exemplary and clear, but since it deals with selection against HbS, this is perhaps not surprising. In respect of other loci there is nothing like such clear evidence. With characteristic trenchancy, Morton (quoted in Krieger *et al.* 1965) remarked that 'whenever enough generations have elapsed for selection to be detectable through mixture estimates, there will be such uncertainty about the ancestral gene frequencies as to make this evidence of selection altogether worthless'.

Miscellaneous features

Under this heading I shall consider a number of unrelated issues that have not been investigated in the detail accorded to the topics elaborated above, but which might throw some light on human genetic adaptation. What can we infer from the universality of polymorphism, from the occurrence of 'human' polymorphisms in non-human primates, from general interspecific comparisons, from secular change through time, and from incompatibility selection?

Universal polymorphisms

It has been observed that there are a number of polymorphisms, such as the ABO, MN, and Rhesus blood groups, that are polymorphic in almost all human populations. An argument follows from this that such polymorphisms must therefore be adaptive, maintained by selection, otherwise they would more often be monomorphic, as a result of the action of genetic drift. Against this view we might argue that one of the most characteristic features of human social behaviour is the exchange of marriage partners and genes between populations, so that even 'isolates' are rarely completely isolated. This behaviour is characteristic of small-scale societies in the present day (Salzano and Callegari-Jacques 1988) and by inference of hunter–gatherers and tribal societies through time (Smith and Layton 1987), and may help to account for the demonstrable high allelic diversity within human populations, and relatively small between-population differences (Lewontin 1972).

Comparison with other primates

Neel (1975) has made the suggestion, though wryly 'without a rigorous defence', that a polymorphism which is adaptive in one species might well crop up in another. In this light we may view with interest the presence in non-human primates of several polymorphisms which appear to be the same as their counterparts in man. Examples include the ABO and MN blood groups, the sex-linked Xg blood group, and the PTC taste sensitivity polymorphism.

In contrast to the argument of the previous paragraph, comparisons between related species are usually undertaken in order to find adaptive explanations of the differences that the comparison reveals, rather than the similarities (Clutton-Brock and Harvey 1984). Comparison of human and non-human primates has offered insights into morphological and behavioural adaptation in man, for example in respect of testis weight and social organization (Harcourt *et al.* 1981). Comparison of other characters also seems to reveal our adaptation, but we should be somewhat chastened by the fact that for even the most characteristic human traits the adaptive value cannot always be taken for granted. Isaac (1983), for example, gave an account of no fewer than eight different hypotheses put forward to account for bipedalism! There are two points to be made in relation to this: firstly, the adaptive value of a genuinely adaptive trait may be very hard to demonstrate; and secondly, there are legitimate alternative possibilities to adaptation. Some of the latter were considered by Darwin himself, for example the effects of habit, and correlations of growth. Later authors, too, have also criticized the view that for any given feature of morphology or behaviour an adaptive explanation must be forthcoming, if only we look hard enough (Gould and Lewontin 1979).

Change through time

Empirical studies of genetic change through time hold some promise for the future, but with a lack of genuine re-sampling other devices, such as age stratification, have been resorted to. For example, Morris (1973) quoted a study by Glass of gene frequencies in the Dunkers, a religious isolate of German origin whose founders settled in Pennsylvania in 1719, where they were later joined by other migrants. Differences between the Dunkers and present day populations in Germany might have been due to founder effect. When generations within the Dunkers were compared, ABO frequencies were apparently stable from generation to generation, by implication under selective constraint. MN blood group frequencies, however, showed large inter-generational differences which were attributed by Glass to genetic drift, though Roberts later suggested that the magnitude of drift implied would be improbable in a single generation, and suggested selective emigration from the sect as a more plausible cause.

De Vries *et al.* (1979) reported that in 1845 several hundred Dutch farmers emigrated to Surinam, South America. Shortly after their arrival, about half died from yellow fever, and a proportion of the remainder died from yellow fever a few months later. The descendants of the survivors were tested for HLA and 24 other genetic systems and a comparison was made between the descendants' gene frequencies and those of the present day Netherlands population. It was claimed that discrepancies in the frequencies of C3, Gm, HLA-B, and GLO were unlikely to be due solely to drift. All the other polymorphisms tested did not differ significantly between groups.

Vogel (1975) quoted a study by Kellerman on ABO blood group frequency change from the fourteenth century to the present, based on blood grouping bone samples from the charnel houses in Oppenheim and Alken. The trends, which showed a consistent increase through time in blood group O and net reductions in A and B, suggested no clear interpretation. In Vogel's opinion, relaxation of selection (due to plague) against blood group O might be responsible, but he is neither convincing nor convinced that this is an appropriate explanation.

Incompatibility selection

By incompatibility selection I mean the relative advantages and disadvantages accruing to particular offspring genotypes as a consequence of incompatibility between mother and fetus, or in the case of pre-zygotic selection, between egg (or mother's body) and sperm. Such selection has long been known to affect the ABO and Rhesus loci, with a net result of elimination of particular heterozygotes (Cavalli-Sforza and Bodmer 1971). Elementary theory shows that selection against heterozygotes results in unstable polymorphism, with the more common allele rapidly achieving fixation. In spite of this, the polymorphisms concerned appear to be stable, so the inference has been made that some form of balancing selection must be acting to outweigh the incompatibility effects.

As remarked at the beginning of this section, the approaches summarized above are disparate. Together, they significantly add to the list of issues addressed in relation to selection, but I feel on balance that they whet the appetite rather than satisfying it.

Summary and conclusions

In drawing together the various threads of this essay, I have in mind not just to condense and reiterate the main points, though I shall emphasize and gloss some of the more striking observations. Rather, I want to place the review more squarely within the contextual framework of evolutionary biology. Finally, and in brief, I shall suggest how future developments may alter our present understanding.

An acknowledged bias of this essay has been its focus on variation within or between populations. I have scarcely considered the species as a whole. Let me try to redress that to some extent here, with some brief, but I believe telling, references to stabilizing selection.

It has been estimated that approximately 75 per cent of fertilized ova are lost (Gill 1992). Clearly not all this fetal wastage is attributable to normalizing selection on genetic variation, but at least half of all spontaneous abortions and miscarriages are caused by chromosome abnormalities (Gehlerter and Collins 1990). Of these, over 20 per cent are triploidies and tetraploidies, and over 50 per cent trisomies. On top of this, genetic disease has been estimated by Weatherall (1991) to affect between about 40 and 50 births per thousand, through single gene disorders, chromosomal abnormalities, and disorders of complex causation. Stabilizing selection has also been shown clearly to occur in relation to birth weight, with babies of between 7.5 and 8 pounds having optimal survival. Changing scale from the phenotypic to the DNA level, we can cite the impressive observation that molecular evolution is apparently faster at the third codon of a triplet than at the other two (Kimura 1983, p. 94). Among several instances of this phenomenon is a comparison of human, rabbit, mouse, and chicken globin sequences. The importance of this observation is that the third codon is often redundant, and its faster rate of evolution had been predictable under neutral theory long before (Lewontin 1974). For us its significance is in the implication of selective constraint (stabilizing selection) acting on codons in the first and second positions.

I shall now turn to the broader evolutionary context to show why this compelling evidence of adaptation from conservation has attracted relatively little attention beside the struggle to show adaptation through variation. What do we know about evolution? There is general agreement that long-term adaptive evolution takes place through natural selection. Opinions differ as to how big a role natural selection plays in speciation and macroevolution, but these are for the most part differences in the emphasis of selective and chance factors, rather than genuinely alternative processes (Ridley 1985). All biologists, however, are agreed that natural selection is the agency by which phenotypic adaptation is achieved and maintained, and many would find compelling Dawkins's argument that there are no other known contenders (Dawkins 1986). There is far less accord, however, in the debate to explain the function of genetic variation at the population level. This stemmed from antecedent arguments about population structure (see Crow 1987, for a historical account), and blossomed with the demonstration by electrophoresis of universally high rates of genetic polymorphism in all species, even those whose morphology had been constant for millions of years (Lewontin 1974, p. 117). What is all this polymorphism doing? Is it there because (according to the Balance Theory) variation increases individual fitness, either through heterozygote advantage, frequency-dependent selection, or other such devices, or is it there by chance, as a result of drifting frequencies of functionally equivalent alleles (as suggested by Neutral Theory)? This opposition of views has determined the direction of research for more than thirty years, and has turned attention away from the importance of stabilizing selection on the species as a whole, towards a predominant focus on intra- and inter-population variation.

Whilst the Neutralist–Selectionist debate has set the agenda, malaria has dominated the discussion. Malaria stands out like a beacon and a siren. We navigate by it, yet it drags us on to the rocks. There is, of course, no question that malaria is the agent responsible for the most profound and fascinating occurrences of natural selection and adaptation in man. And yet we are tired of it, at least Lewontin is (Lewontin 1974, p. 199). The reason for this is that too much has been asked of it: malaria and the genetic adaptations it evinces are continually invoked as a model for selection by other agents on other loci, but nothing else has been able to measure up. I think we can safely say that malaria is not a good model for explanation of other polymorphisms. If there were other agents exerting this kind of influence they would by now surely have been found. Malaria is fascinating in its own right, continually stimulating new discoveries such as the recently revealed HLA associations, but we should recognize that it is *the* case rather than *an* example.

There may be only a quantitative difference, and no theoretical one, between the effect of malaria and other agents on polymorphisms, but in practical terms other contenders have all been outside of the circle of confidence, where even a sceptic has to admit heterozygote advantage, and on the ambiguous ground where we are forced feebly to say that selection could have done this, but so too could chance. The attribution of distributions to chance seems to me problematic, because we can calculate the distribution of random probabilities, but have to assign a cause in individual cases. This can result in the plea that we should expect a result like this by chance every once in a while. Perhaps in interpreting the population frequency of any one recessive disease, the distribution of others should be taken into account.

To move on from malaria and recessive disease, perhaps the greatest disappointment to me in the material reviewed above is the overwhelming problem of sample size in demographic studies. The demographic approach has great appeal as the ultimate proof of the pudding, but the numerical constraints have prevented us from tasting success. Perhaps the approach most likely to succeed in this regard would be a comparison of mother–offspring genotypes (Manly 1985; Endler 1986) since this would circumvent the problems of low confidence of paternity mentioned above. Such an initiative would not be difficult to design, but the organization and execution would require a collective effort of will (and funding) on a scale which has so far eluded selection studies.

Of the other methods, formal models are ingenious and (in contrast to demographic models) adequate data can readily be gathered to test them. Already they suggest a degree of balancing selection in the maintenance of polymorphism, and as the models are refined and more data made available (especially at the DNA level, e.g. Bowcock *et al.* 1991), they will yield increasingly authoritative results. Recent work on the distribution of genes in relation to linguistics suggests a predominant role for settlement migration, yet some of the more comprehensive surveys (such as Sokal 1989) speak for multiple causation rather than a single determinant.

For the future, studies at the molecular level will unquestionably bring insights, but understanding will not follow just because a new technique is in place, as the history of electrophoresis shows (Crow 1987). Areas in which molecular anthropology ought to benefit us in the search for selection include the following. Firstly, genotypes will be distinguished at the finest level, with the inevitable result that apparently homogeneous mutations will be revealed as heterogeneous. This has already happened in respect of a number of traits, for example HbS and the CF gene. Confidence of allelic individuality will confer authority to judgements of the relative probabilities that allele frequencies have been achieved by chance or through selection. Secondly, the linkage groups of markers will be explored much more fully, with the result that physiological and biochemical functions of haplotypes will be increasingly understood, as in the case of hereditary ovalocytosis. Thirdly, it will be possible systematically to compare the extent of variation between coding and non-coding DNA regions.

At present, if we are asked what constitutes the evidence that genetic variation within mankind is adaptive, we have to answer cautiously. To be sure there are a few unambiguous cases, for example body shape, the malarial resistance polymorphisms, and persistent adult intestinal lactase. If pressed to answer for genetic variation in general, we can point to widespread differences in allelic function, to intriguingly high levels of recessive disease in many populations, to deviations from models based on neutral alleles, and to the possibility of reasonable levels of selection going undetected. But all this adds up to no more than a 'maybe'. The uncomfortable fact remains that, as many authorities have pointed out, selection coefficients sufficient to direct evolution can be below the resolution of our methods of detection, and the evidence they leave behind is ambiguous. This is immensely frustrating, and may explain why so few recent studies address the question of selection at all. Yet another deterrent may be the not implausible inference that polymorphism has been subject to selection in the past, but is no longer so affected in the present. This really is an intellectually unsatisfactory conclusion, an acknowledgement that present day distributions may reflect selective forces which can be conjectured but not demonstrated by function or fitness.

If the question is re-phrased to ask is mankind's genetic make-up the result of adaptation through natural selection, then our answer changes from caution to certainty. As well as the evidence from genetic polymorphism within populations, we have the much more compelling examples of selection acting to eliminate variation beyond the normal range. In Darwin's metaphor, 'natural selection is daily and hourly scrutinising, throughout the world, the slightest variations' (Darwin 1859, Penguin edition, 1968, p. 133). No biologist can seriously doubt the truth of this insight, but from this survey of research into human genetic variation the evidence of selection's scrutiny is to be found more in what we have in common than in our differences.

References

Abel, L., Demenais, F., Prata, A., Souza, A.E., and Dessein, A. (1991). Evidence for the segregation of a major gene in human susceptibility/resistance to infection by *Schistosoma mansoni*. *Am. J. Hum. Genet.*, **48**, 959–70.

Agarwal, D.P. and Goedde, H.W. (1991). Human variability in alcohol and aldehyde metabolism: vulnerability to toxic effects. In *Ecogenetics* (ed. P. Grandjean), pp. 141–58. Chapman and Hall, London.

Aird, I., Bentall, H.H., and Roberts, J.A.F. (1953). A relationship between cancer of stomach and ABO blood groups. *Brit. Med. J.*, **1**, 799–801.

Allison, A.C. (1954). The distribution of the sickle-cell trait in East Africa and elsewhere and its apparent relationship to the incidence of subtertian malaria. *Trans. R. Soc. Trop. Med. Hyg.*, **48**, 312–18.

Amato, D. and Booth, P.B. (1977). Hereditary ovalocytosis in Melanesians. *Papua New Guinea Med. J.*, **20**, 26.

Averill, B.K. (1980). Detection of natural selection by methods based on models of the behaviour of neutral alleles. *J. Biosoc. Sci.*, **12**, 469–79.

Baker, P.T. (1988). Human adaptability. In *Human biology* (ed. G.A. Harrison, J.M. Tanner, D.R. Pilbeam, and P.T. Baker), pp. 430–547. Oxford University Press.

Barbujani, G. and Sokal, R.R. (1990). Zones of sharp genetic change in Europe are also linguistic boundaries. *Proc. Natl. Acad. Sci.*, **87**, 1816–19.

Barbujani, G. and Sokal, R.R. (1991). Genetic population structure of Italy. II. Physical and cultural barriers to gene flow. *Am. J. Hum. Genet.*, **48**, 398–411.

Barbujani, G., Vian, P., and Fabbris, L. (1992). Cultural barriers associated with large gene frequency differences among Italian populations. *Hum. Biol.*, **64**, 479–95.

Beet, E.A. (1946). Sickle cell disease in the Balovale District of Northern Rhodesia. *East Af. Med. J.*, **23**, 75–86.

Bertranpetit, J. and Cavalli-Sforza, L.L. (1991). A genetic reconstruction of the history of the population of the Iberian Peninsula. *Ann. Hum. Genet.*, **55**, 51–67.

Bittles, A.H. and Smith, M.T. (1990). ABO and Rh(D) blood group frequencies in the Ards Peninsula, Northern Ireland: evidence for the continuing existence of a major politico-religious boundary. *Ann. Hum. Biol.*, **18**, 253–8.

Blackman, G.E., Caesar, A.A.L., Gullick, C.F.W.R., Steers, J.A., and Worswick, C.D.N. (eds.) (1963). The atlas of Britain and Northern Ireland. Oxford University Press.

Bodmer, W.F. (1968). Demographic approaches to the measurement of differential selection in human populations. *Proc. Natl. Acad. Sci.*, **59**, 690–9.

Bodmer, W.F. (1973). Population studies and the measurement of natural selection, with special reference to the HL-A system. *Isr. J. Med. Sci.*, **9**, 1503–18.

Bodmer, W.F. and Cavalli-Sforza, L.L. (1976). *Genetics, evolution and man*. W.H. Freeman, San Franciso.

Bodmer, W.F., Trowsdale, J., Young, J., and Bodmer, J. (1988). Gene clusters and the evolution of the major histocompatibility system. *Philos. Trans. R. Soc. London*, **B319**, 615–29.

Boutté, M.I. (1987). 'The Stumbling Disease': a case study of stigma among Azorean–Portuguese. *Soc. Sci. Med.*, **24**, 209–17.

Bowcock, A.M., Kidd, J.R., Mountain, J.L., Hebert, J.M., Carotenuto, L., Kidd, K.K., and Cavalli-Sforza, L.L. (1991). Drift, admixture and selection in human evolution: a study with DNA polymorphisms. *Proc. Natl. Acad. Sci. USA*, **88**, 839–43.

Boyd, W.C. (1973). Four achievements of the genetical method in physical anthropology. In *Human populations, genetic variation and evolution* (ed. L.N. Morris), pp. 20–31. Intertext Books, Aylesbury.

Brown, E.S. (1965). Distribution of ABO and Rhesus (D) blood groups in the north of Scotland. *Heredity*, **20**, 289–303.

Buetler, E. (1989). Glucose-6-phosphate dehydrogenase: New perspectives. *Blood*, **73**, 1397–401.

Calabrese, E.J. (1984). *Ecogenetics: Genetic variation in susceptibility to environmental agents.* Wiley Interscience, New York.

Calabrese, E.J. (1991). Genetic predisposition to ocupationally-related diseases: current status and future directions. In *Ecogenetics* (ed. P. Grandjean), pp. 21–56. Chapman and Hall, London.

Caro, T. and Borgerhoff Mulder, M. (1987). The problem of adaptation in the study of human behaviour. *Ethology and Sociobiology*, **8**, 61–72.

Cartwright, R.A., Glashan, R.W., Rogers, H.J., Ahmad, R.A., Barham-Hall, D., Higgins, E., and Kahn, M.A. (1982). Role of *N*-acetyltransferase phenotypes in bladder carcinogenesis. A pharmacogenetic epidemiological approach to bladder cancer. *Lancet*, **11**, 842–5.

Casals, T., Vázquez, C., Lázaro, C., Girbau, E., Giménez, F.J., and Estivel, X. (1992). Cystic fibrosis in the Basque Country: High frequency of mutation ΔF508 in patients of Basque origin. *Am. J. Hum. Genet.*, **50**, 404–10.

Cavalli-Sforza, L.L. (1991). Genes, peoples, and languages. *Scientific American*, **265** (5), 72–8.

Cavalli-Sforza, L.L. and Bodmer, W.F. (1971). *The Genetics of Human Populations.* W.H. Freeman, San Franciso.

Cavalli-Sforza, L.L., Piazza, A., Menozzi, P., and Mountain, J. (1988). Reconstruction of human evolution: bringing together genetic, archaeological, and linguistic data. *Proc. Natl. Acad. Sci. USA*, **85**, 6002–6.

Cavalli-Sforza, L.L., Minch, E., and Mountain, J.L. (1992). Coevolution of genes and languages revisited. *Proc. Natl. Acad. Sci. USA*, **89**, 5620–4.

Clarke, B. (1975). Frequency-dependent and density-dependent natural selection. In *The role of natural selection in human evolution* (ed. F.M. Salzano), pp. 187–200. North-Holland, Amsterdam.

Clutton-Brock, T.H. and Harvey, P.H. (1984). Comparative approaches to investigating adaptation In *Behavioural ecology, an evolutionary approach*, (2nd edn) (ed. J.R. Krebs and N.B. Davies), pp. 7–29. Blackwell, Oxford.

Conneally, P.M., Merritt, A.D., and Pao-Lo, Y. (1973). Cystic fibrosis: population genetics. *Texas Rep. Bio. Med.*, **31**, 639–50.

Cook, G.C. (1978). Did persistence of intestinal lactase into adult life originate on the Arabian Peninsula? *Man, n.s.*, **13**, 418–27.

Crawford, M.H., Mielke, J.H., Devor, E.J., Dykes, D.D., and Polesky, H.F.

(1981). Population structure of Alaskan and Siberian indigenous communities. *Am. J. Phys. Anth.*, **55**, 167–85.
Crow, J.F. (1958). Some possibilities for the analysis of selection intensities in man. *Hum. Biol.*, **30**, 1–13.
Crow, J.F. (1987). Muller, Dobzhansky, and overdominance. *J. Hist. Biol.*, **20**, 351–80.
Crow, J.F. (1989) Update to 'Some possibilities for the analysis of selection intensities in man'. *Hum. Biol.*, **61**, 776–80.
Danks, D.M., Allan, J., and Anderson, C.M. (1965). A genetic study of fibrocystic disease of the pancreas. *Ann. Hum. Genet.*, **28**, 323–55.
Darwin, C.R. (1859). *The origin of species* (1st edn). Penguin, Harmondsworth, 1968.
Darwin, C.R. (1874). *The descent of man and selection in relation to sex* (2nd edn). John Murray, London.
Dawkins, R. (1976). *The selfish gene*. Oxford University Press.
Dawkins, R. (1986). *The blind watchmaker*. Longman Scientific and Technical, Harlow.
De Braekeleer, M. and Daigneault, J. (1992). Spatial distribution of the DF508 mutation in cystic fibrosis: a review. *Hum. Biol.*, **64**, 167–74.
De Vries, R.R.P., Meera Khan, P., Bernini, L.F., Van Loghem, E., and Van Rood, J.J. (1979). Genetic control of survival in epidemics. *J. Immunogenetics*, **6**, 271–87.
Desmond, A. and Moore, J. (1991). *Darwin*. Michael Joseph, London.
Diamond, J. (1991). *The rise and fall of the third chimpanzee*. Radius, London.
Dobzhansky, T. (1972). *Natural selection in mankind*. In *The structure of human populations* (ed. G.A. Harrison and A.J. Boyce). Clarendon, Oxford.
Durham, W.H. (1983). Testing the malaria hypothesis in West Africa. In *The distribution and evolution of hemoglobin and globin loci* (ed. J.E. Bowman). Elsevier, Amsterdam.
Endler, J.A. (1986). *Natural selection in the wild*. Princeton University Press.
Fishberg, M. (1911). *The Jews, a study of race and environment*. Scribner, New York.
Fisher, R.A. (1930). The distribution of gene ratios for rare mutations. *Proc. R. Soc. Edinburgh*, **50**, 205–20.
Fix, A.G. and Lie-Injo, L.E. (1975). Genetic microdifferentiation in the Semai Senoi of Malaysia. *Am. J. Phys. Anth.*, **43**, 47–56.
Flatz, G. (1967). Haemoglobin E: distribution and population dynamics. *Humangenetik*, **3**, 189–234.
Flint, J., Hill, A.V.S., Bowden, D.K., Oppenheimer, S.J., Sill, P.R., Serjeantson, S.W., Bana-Koiri, J., Bhatia, K., Alpers, M.P., Boyce, A.J., Weatherall, D.J., and Clegg, J.B. (1986). High frequencies of α-thalassaemia are the result of natural selection by malaria. *Nature*, **321**, 774–50.
Ford, E.B. (1964). *Ecological genetics*. (1st edn). Methuen, London.
Ford, E.B. (1975). *Ecological genetics*. (4th edn). Chapman and Hall, London.
Fox, C.F. (1943). *The personality of Britain*. (4th edn). National Museum of Wales, Cardiff.
Gehlerter, T.D. and Collins, F.S. (1990). *Principles of medical genetics*. Williams and Wilkins, Baltimore.
Gill, T.J. (1992). Influence of MHC and MHC-linked genes on reproduction. *Am. J. Hum. Genet.*, **50**, 1–5.

Gonzalez, R., Ballester, J.M., Estrada, M., Lima, F., Martinez, G., Wade, M., Colombo, B., and Vento, R. (1976). A study of the genetical structure of the Cuban population: red cell and serum biochemical markers. *Am. J. Hum. Genet.*, **28**, 585–96.

Gould, S.J. (1991). Male nipples and clitoral ripples. In *Bully for brontosaurus*, pp. 124–38. Penguin Books, Harmondsworth.

Gould, S.J. and Lewontin, R.C. (1979). The spandrels of San Marco and the Panglossian paradigm: a critique of the adaptationist programme. *Proc. R. Soc. London*, **B205**, 581–98.

Grandjean, P. (ed.) (1991). *Ecogenetics*, pp. 21–56. Chapman and Hall, London.

Haldane, J.B.S. (1940). The blood-group frequencies of European peoples, and racial origins. *Hum. Biol.*, **12**, 457–80.

Haldane, J.B.S. (1949). Disease and evolution. *La Ricerca Sci. Suppl.*, **19**, 3–11.

Hamilton, W.D. (1964). The evolution of social behaviour. *J. Theor. Biol.*, **7**, 1–52.

Harada, S., Agarwal, D.P., and Goedde, H.W. (1985). Aldehyde dehydrogenase polymorphism and alcohol metabolism in alcoholics. *Alcohol*, **2**, 391–2.

Harcourt, A.H., Harvey, P.H., Larson, S.G., and Short, R.V. (1981). Testis weight, body weight and breeding system in primates. *Nature*, **293**, 55–7.

Harris, H. (1975). *The principles of human biochemical genetics*. North-Holland, Amsterdam.

Harrison, G.A. (1988). Human genetics and variation. In *Human biology* (ed. G.A. Harrison, J.M. Tanner, D.R. Pilbeam, and P.T. Baker), pp. 145–336. Oxford University Press.

Hartl, D.L. and Clark, A.G. (1989). *Principles of population genetics* (2nd edn). Sinauer Associates, Sunderland, Massachusetts.

Hendrick, P.W. and Thomson, G. (1983). Evidence for balancing selection at HLA. *Genetics*, **104**, 449–56.

Hill, A.V.S., Elvin, J., Willis, A.C., Aidoo, M., Allsop, C.E.M., Gotch, F.M., *et al.* (1992). Molecular analysis of the association of HLA-B53 and resistance to severe malaria. *Nature*, **360**, 434–9.

Hiorns, R.W. (1972). The dynamics of migration and selection in human populations. In *The mathematical theory of the dynamics of biological populations* (ed. M.S. Bartlett and R.W. Hiorns), pp. 203–22. Academic Press, London.

Hiorns, R.W. and Harrison, G.A. (1970). Sampling for the detection of natural selection by age group genetic differences. *Hum. Biol.*, **42**, 53–64.

Isaac, G.L. (1983). Aspects of human evolution. In *Evolution from molecules to man* (ed. D.S. Bendall). Cambridge University Press.

Johnson, R.C., Cole, R.E., and Ahern, F.M. (1981). Genetic interpretation of racial/ethnic differences in lactose absorption and tolerance: a review. *Hum. Biol.*, **53**, 1–13.

Jorde, L.B. (1985). Human genetic distance studies. *Ann. Rev. Anth.*, **14**, 343–73.

Jorde, L.B. and Durbize, P. (1986). Opportunity for selection in Utah Mormons. *Hum. Biol.*, **58**, 97–114.

Jorde, L.B. and Lathrop, G.M. (1988). A test of the heterozygote-advantage hypothesis in cystic fibrosis. *Am. J. Hum. Genet.*, **42**, 808–15.

Jorde, L.B., Pitkänen, K.J., O'Brien, E., and Eriksson, A.W. (1992). Consanguinity and genetic disease in Finland's Swedish-speaking minority. In *Minority populations, genetics, demography and health* (ed. A.H. Bittles and D.F. Roberts), pp. 14–34. Macmillan, Basingstoke and The Galton Institute.

Kalow, W. (1991). Lessons from pharmacogenetics. In *Ecogenetics* (ed. P. Grandjean), pp. 125–40. Chapman and Hall, London.

Kidd, K.K. (1987). Population genetics of a disease. *Nature*, **327**, 287–8.

Kimura, M. (1983). *The neutral theory of molecular evolution*. Cambridge University Press.

Knudson, A.G., Wayne, L., and Hallett, W.Y. (1967). On the selective advantage of cystic fibrosis heterozygotes. *Am. J. Hum. Genet.*, **19**, 388–92.

Kopeć, A.C. (1970). *The distribution of the blood groups in the United Kingdom*. Oxford University Press.

Kretchmer, N.O. (1972). Lactose and lactase. In *Human physiology and the environment in health and disease. Readings from Scientific American* (ed. A.J. Vander), pp. 47–54. W.H. Freeman, San Franciso.

Krieger, H., Morton, N.E., Mi, M.P., Azevêdo, E., Freire-Maia, A., and Yasuda, N. (1965). Racial admixture in north-eastern Brazil. *Ann. Hum. Genet.*, **29**, 113–25.

Lewontin, R.C. (1972). The appportionment of human diversity. *Evol. Biol.*, **6**, 381–98.

Lewontin, R.C. (1974). *The genetic basis of evolutionary change*. Columbia University Press.

Lewontin, R.C. and Krakauer, L. (1973). Distribution of gene frequency as a test of the theory of selective neutrality of polymorphisms. *Genetics*, **74**, 175–95.

Li, C.C. (1976). *First course in population genetics*. The Boxwood Press, Pacific Grove, California.

Lie-Injo, L.E. (1976). Genetic relationships of several aboriginal groups in South East Asia. In *The origin of the Australians* (ed. R.I. Kirk and A.G. Thorne), pp. 277–306. Australian Institute of Aboriginal Studies, Canberra.

Livingstone, F.B. (1983). The malaria hypothesis. In *The distribution and evolution of hemoglobin and globin loci* (ed. J.E. Bowman), pp. 15–44. Elsevier, Amsterdam.

Macintyre, S. and Sooman, A. (1991). Non-paternity and prenatal genetic screening. *Lancet*, **338**, 869–71.

Manly, B.F.J. (1985). *The statistics of natural selection*. Chapman and Hall, London.

Mange, A.P. and Mange, E.J. (1990). *Genetics: Human aspects*. Sinauer Associates, Sunderland, Massachusetts.

Malhotra, K.C. (1978). Natural selection and colour blindness: fresh data on Indian castes. *Genet. Res. Camb.*, **31**, 203–7.

Mayr, E. (1963). *Animal species and evolution*. Bellknap Press of Harvard University, Cambridge, Massachusetts.

Mayr, E. (1970). *Populations, species, and evolution*. Bellknap Press of Harvard University, Cambridge, Massachusetts.

McClellan, T., Jorde, L.B., and Skolnick, M.H. (1984). Genetic distances between the Utah Mormons and related populations. *Am. J. Hum. Genet.*, **36**, 836–57.

McCluer, J.W. (1978). On the probability of demonstrating differential fertility in genetic studies. *Ann. Hum. Genet., London*, **42**, 59–75.

McKusick, V.A. (1979). Nonhomogeneous distribution of recessive diseases. In *Genetic diseases among Ashkenazi Jews* (ed. R.M. Goodman and A.G. Motulsky), pp. 271–84. Raven Press, New York.

Meindl, R.S. (1987). Hypothesis: a selective advantage for cystic fibrosis heterozygotes. *Am. J. Phys. Anth.*, **74**, 39–45.

Miller, L.H., Mason, S.J., Clyde, D.F., and McGinniss, M.H. (1976). The resistance factor to *Plasmodium vivax* in blacks. *New England J. Med.*, **295**, 302–4.

Morris, L.N. (ed.) (1973). *Human populations, genetic variation and evolution.* Intertext Books, Aylesbury.

Morton, N.E. (1970). Genetic structure of northeastern Brazilian populations. In *The ongoing evolution of Latin American populations* (ed. F.M. Salzano). Charles C. Thomas, Springfield, Illinois.

Morton, N.E. (1975). Kinship, fitness and evolution. In *The role of natural selection in human evolution* (ed. F.M. Salzano), pp. 133–54. North-Holland, Amsterdam.

Morton, N.E. and Chung, C.S. (1959). Are the MN blood groups maintained by selection? *Am. J. Hum. Genet.*, **11**, 237–51.

Morton, N.E., Krieger, H., and Mi, M.P. (1966). Natural selection on polymorphisms in northeastern Brazil. *Am. J. Hum. Genet.*, **18**, 153–71.

Motulsky, A.G. (1979). Possible selective effects of urbanization on Askenazi Jews. In *Genetic diseases among Ashkenazi Jews* (ed. R.M. Goodman and A.G. Motulsky), pp. 301–14. Raven Press, New York.

Motulsky, A.G. (1989). Update to 'Metabolic polymorphisms and the role of infectious diseases in human evolution'. *Hum. Biol.*, **61**, 870–7.

Mourant, A.E., Kopec, A.C., and Dominiewska-Sobjzak, K. (1976). *The distribution of the human blood groups and other polymorphisms* (2nd edn). Oxford University Press.

Mourant, A.E., Kopec, A.C., and Dominiewska-Sobjzak, K. (1978). *Blood groups and diseases.* Oxford University Press.

Myrianthopolous, N.C. and Aronson, S.M. (1966). Population dynamics of Tay-Sachs disease. I. Reproductive fitness and selection. *Am. J. Hum. Genet.*, **18**, 313–27.

Nagel, R.L. and Roth, E.F. (1989). Malaria and red cell genetic defects. *Blood*, **74**, 1213–21.

Neel, J.V. (1975). The study of 'natural' selection in man: last chance. In *The role of natural selection in human evolution* (ed. F.M. Salzano), pp. 355–68. North-Holland, Amsterdam.

Neel, J.V. (1979). History and the Tay-Sachs allele. In *Genetic diseases among Ashkenazi Jews* (ed. R.M. Goodman and A.G. Motulsky), pp. 285–99. Raven Press, New York.

Neel, J.V. (1989). Update to 'The study natural selection in primitive and civilized human population'. *Hum. Biol.*, **61**, 811–23.

Neel, J.V. and Schull, W.J. (1968). On some trends in understanding the genetics of man. *Persp. Biol. Med.*, **11**, 565–602.

Nei, M. (1988). Relative roles of mutation and selection in the maintenance of genetic variability. *Philos. Trans. R. Soc. London*, **B319**, 615–29.

Nei, M. and Yokoyama, S. (1981). Estimation of fitness reduction due to a chronic disease in man. *Ann. Hum. Genet.*, **45**, 261–5.

O'Brien, E., Jorde, L.B., Rönnlöf, B., Fellman, J.O., and Eriksson, A.W. (1988). Founder effect and genetic disease in Sottunga, Finland. *Am. J. Phys. Anth.*, **77**, 335–46.

Post, R.H. (1962). Population differences in red and green color vision deficiency. *Eugenics Quarterly*, **9**, 131–46; **10**, 84–5.

Race, R.R. and Sanger, R. (1975). *Blood groups in man* (6th edn). Blackwell Scientific Publications, Oxford.

Reed, T.E. (1975). Selection and the blood group polymorphisms. In *The role of natural selection in human evolution* (ed. F.M. Salzano), pp. 231–46. North-Holland, Amsterdam.

Ridley, M. (1985). *The problems of evolution*. Oxford University Press.

Roberts, J.A.F. (1953). An analysis of the blood-group records of the north of England. *Heredity*, **7**, 361–88.

Roberts, D.F. (1973*a*). *Climate and human variability*, Module in Anthropology No. 34. Addison-Wesley, Reading, Massachusetts.

Roberts, D.F. (1973*b*). The origins of genetic variation in Britain. In *Genetic variation in Britain* (ed. D.F. Roberts and E. Sunderland), pp. 1–16. Taylor and Francis, London.

Robins, A.H. (1991). *Biological perspectives on human pigmentation*. Cambridge University Press.

Russell, J.C. (1948). *British medieval population*. University of New Mexico Press, Albuquerque.

Ryder, L.P., Svejgaard, A., and Dausset, J. (1981). Genetics of HLA disease association. *Ann. Rev. Genet.*, **15**, 169–88.

Salzano, F.M. and Callegari-Jacques, S.M. (1988). *South American Indians: a case study in evolution*. Clarendon Press, Oxford.

Saugstaad, L.F. (1975). Anthropological significance of phenylketonuria. *Clin. Genet.*, **7**, 52–61.

Sawchuck, L.A. and Herring, D.A. (1984). Respiratory tuberculosis mortality among the Sephardic Jews of Gibraltar. *Hum. Biol.*, **56**, 291–306.

Serjeant, G.R. (1992). *Sickle cell disease* (2nd edn). Oxford University Press.

Serjeantson, S., Bryson, K., Amato, D., and Babona, D. (1977). Malaria and hereditary ovalocytosis. *Hum. Genet.*, **37**, 161–7.

Smith, M.T. (1989). Prospects for the study of genetic demography in medieval English populations. *Actes des 4èmes Journées Anthropologiques*, pp. 95–106. Editions du CRNS, Paris.

Smith, M.T. and Layton, R.H. (1987). Why Haven't We Speciated? *New Community*, **13**, 367–72.

Sokal, R.R. (1989). Update to Haldane's 'Blood-group frequencies of European peoples and racial origins'. *Hum. Biol.*, **61**, 691–702.

Sokal, R.R., Oden, N.L., and Thomson, B.A. (1988). Genetic changes across language boundaries in Europe. *Am. J. Phys. Anth.*, **76**, 337–61.

Spickett, S.G. (1962). Genetics and the epidemiology of leprosy. *Leprosy Rev.*, **33**, 76–93.

Spuhler, J.N. (1976). The maximum opportunity for natural selection in some human populations. In *Demographic anthropology* (ed. E.B.W. Zubrow). University of New Mexico Press, Albuquerque.

Stewart, F.M. (1977). Computer algorithm for obtaining a random set of allele frequencies for a locus in an equilibrium population. *Genetics*, **86**, 482–3.

Thomson, G. (1988). HLA disease associations: models for insulin dependent diabetes mellitus and the study of complex human genetic disorders. *Ann. Rev. Genet.*, 22, 31–50.

Veillette, S., Perron, M., Mathieu, J., Prévost, C., and Hébert, G. (1992). Socio-cultural factors influencing the spread of myotonic dystrophy in the Saguenay-Lac-Saint-Jean region of the Province of Quebec. In *Minority populations, genetics, demography and health* (ed. A.H. Bittles and D.F. Roberts), pp. 83–101. Macmillan, Basingstoke and The Galton Institute.

Vogel, F. (1975). ABO blood groups, the HL-A system and diseases. In *The role of natural selection in human evolution* (ed. F.M. Salzano), pp. 247–69. North-Holland, Amsterdam.

Vogel, F., Pettenkoffer, H.J., and Helmbold, W. (1960). Uber die populations-genetik der ABO-blutgruppen. *Acta Genet. Statist. Med.*, 10, 267–94.

Wainscoat, J.S., Hill, A.V.S., Boyce, A.J., Flint, J., Hernandez, M., Thein, S.L., *et al.* (1986). Evolutionary relationships of human populations from an analysis of nuclear DNA polymorphism. *Nature*, 319, 491–3.

Wallace, B. (1981). *Basic population genetics*. Columbia University Press, New York.

Watkin, I.M. (1960). A Viking settlement in Little England beyond Wales: ABO blood group evidence. *Man*, 60, 148–53.

Watkin, I.M. (1986). ABO blood group distribution in Wales in relation to human settlement. In *Genetic and population studies in Wales* (ed. P.S. Harper and E. Sunderland), pp. 118–46. University of Wales Press, Cardiff.

Weatherall, D.J. (1991). *The new genetics and clinical practice* (3rd edn). Oxford University Press.

Williams, G.C. (1966). *Adaptation and natural selection*. Princeton University Press.

Williams, E.M. (1986). Genetic studies of Welsh Gypsies. In *Genetic and population studies in Wales* (ed. P.S. Harper and E. Sunderland), pp. 186–211. University of Wales Press, Cardiff.

Williams, E.M. and Harper, P.S. (1977). Genetic study of Welsh Gypsies. *J. Med. Genet.*, 14, 172–76.

Woolf, C.M. and Dukepoo, F.C. (1973). Hopi Indians, inbreeding and albinism. In *Human populations, genetic variation and evolution* (ed. L.N. Morris), pp. 392–408. Intertext Books, Aylesbury.

Woolhouse, M.E.J. (1992). Evidence for genetic factors for resistence/susceptibility to schistosome infection. *Am. J. Hum. Genet.*, 51, 206–7.

Wright, S. (1943). Isolation by distance. *Genetics*, 28, 114–38.

Zlotogora, J., Ziegler, M., and Bach, G. (1988). Selection in favor of lysosomal storage disorders? *Am. J. Hum. Genet.*, 42, 271–3.

2
PHYSIOLOGICAL ADAPTATION

G. Ainsworth Harrison

Introduction

This contribution is concerned with the changes which occur in the individual human organism when it experiences new environments. Many of these changes, though by no means all, facilitate living in the new environment and are recognized as components of 'individual adaptability'. Much of the ecological success of the human species is attributed to this adaptability, which is exceptional among animals and has played a major role in allowing our species to occupy every kind of environment on Earth. The unique features of human adaptability lie in behaviour, which of course is also at the root of technology. But it also appears that human beings are outstandingly adaptable in other physiological systems, and it is these to which attention will be given here. Although 'individual adaptability' is a species characteristic, i.e. all individuals possess the capacity, the degree and effectiveness of that capacity may vary from person to person. Some of this variation is probably genetically determined—indeed it needs to have been in the past to have evolved, by natural selection, as a species characteristic—but it is usually taken that when two individuals differ in the state of a physiological system affecting adaptability it is because they are in, or have in the past experienced, different environments.

Adaptability is demonstrated in most if not all physiological systems: cardio-vascular, respiratory, haemopoeitic, endocrine, muscular along with neural. It also occurs at the cellular and tissue levels as well as at the system level. Less obviously, components differ in their efficiency and response time. Usually responses that can be quickly elicited—alarm responses—cannot be maintained for long: they are biologically costly and involve risk. Typically they are replaced by more economic processes which can be long sustained but which, however, also take time to develop.

The continuum between immediate and long term physiological responses, can be further extended to include developmental processes, including growth. Growth is clearly only modifiable during the period that it takes place and phenotypes formed by environments during growth will persist in adults whether appropriate to their environments or not.

But growth adaptability is the most economical component of individual adaptability, requiring no active specific maintenance in the adult, and most mammals, at least, tend to live their whole lives in the kinds of environment in which they developed.

At the other extreme from alarm response in the flexibility versus economy trade-offs, is genetic adaptation. Here the phenotypes are fixed in the DNA and are independent of the immediate environment, following throughout life a more or less unmodifiable course. Because of this, genetic adaptation is not a component of individual adaptability, except in so far as variation in such adaptability may be genetically determined.

Climate

The model of individual adaptability has been most precisely developed in relation to climatic adaptability, and the physiological mechanisms in this adaptability have been most thoroughly worked out. Climatic adaptability encompasses thermal elements which affect heat exchanges between the organism and its environment—and specifically what happens when temperate man moves to tropical or arctic conditions (reviewed in Baker 1988); but it is also evidenced in responses to changed barometric pressure, particularly the consequences of living at high altitude environments (reviewed in Baker 1978, 1988), and in the responses to high or low levels of ultraviolet radiation (reviewed in Robins 1991).

The alarm responses on being exposed to heat mainly involve the cardio-vascular and sweating systems. General peripheral vasodilation and increased heart rate bring more blood to the skin, raise skin temperature, and thereby increase heat loss through convection and radiation to the environment (or reduce the amount gained from the environment). Evaporation of water from the skin leads to loss of heat through the energy required for the latent heat of vaporization of water, and is most effective in low humidity conditions with air flow moving saturated air which would otherwise envelope the body. Alarm responses to the cold are in part the converse, with peripheral vasoconstriction leading to a fall in skin temperature and thus a reduction in the difference between skin and air temperature. Shivering is another alarm reaction.

There is much evidence that adaptation to a major temperature change improves with length of exposure. Migrants from temperate conditions begin to be able to work at temperatures they initially found to be intolerably high, and the likelihood of heat stroke progressively diminishes. So far as the cold is concerned migrants become less susceptible to frost-bite and, among other indicators of performance, sleep more easily. These are the well known phenomena of thermal acclimatization.

Acclimatization may continue to improve over considerable time, but most people appear to achieve most of their improvement over a 14–21 day period, depending on various aspects of lifestyle. Behavioural responses play a very significant role, especially in cold acclimatization, but other systems are involved. Basal metabolic rate, for example, is lowered in the heat and raised in the cold, and much of heat acclimatization depends upon facilitation of the sweating mechanism. The changes associated with acclimatization are more efficient than the immediate responses, especially in reducing the strain on the cardio-vascular system.

Growth and physique appear to be little affected by the immediate thermal environment of human individuals, but this is not the case for other mammals, e.g. rats and mice. There is also evidence that the number of human sweat glands that are active is dependent upon early childhood thermal experiences and such long term effects may be quite common in various adaptable mechanisms.

Although temperature does not directly affect physique development, there is a strong relationship between geographical variation in thermal conditions and physique. This relationship arises from genetic differences produced by natural selection. Either by selection for overall size, or for shape (or both), one finds that native tropical peoples are characterized by having high surface areas in relation to their volume, which facilitates heat loss, while groups that have evolved in cold environments tend to have relatively low surface areas.

Similar categories of adaptation are to be found with respect to altitude and especially with respect to the hypoxia experienced in high mountains. Alarm responses quickly develop in those transported to high altitude, involving increased respiratory and cardiac rates as the body attempts to compensate for the reduced partial pressure of oxygen. Again acclimatization occurs with time as a variety of more efficient respiratory and cardio-vascular changes occur. Blood changes include an increase in haemoglobin concentration and a shift in the haemoglobin/oxygen dissociation curve. Myoglobin levels in muscle increase and various intracellular enzyme changes also appear to occur in oxygen deprived tissues.

There is evidence from studies of recent migrants to and from high altitude zones that growth and development too are affected by the immediate environment. Hypoxia is associated with low birth weight but high placental weight, slow growth, late menarche, and hypertrophy of the thorax, which increases vital capacity. The typical 'barrel-shaped' thorax of many mountain peoples, however, is partly genetically fixed since it is also found in the offspring of native highlanders who grow up in lowland conditions. There is evidence for other genetic adaptations. High altitude adaptation thus shows all four categories of response, alarm, acclimatization, developmental, and genetic.

Responses to ultraviolet radiation appear to fall primarily in the second and fourth category. Tanning and epidermal thickening can be thought of as 'acclimatizations' to increased radiation, since they take some days to develop appreciably. Skin colour differences between the world's populations have probably mainly arisen through natural selection for varying amounts of melanocyte activity irrespective of immediate environmental stimulation.

An interesting question raised by the tanning process is whether there exists a physiological memory for acclimatization, i.e. whether acclimatization occurs more rapidly or effectively on second and subsequent exposures. In tanning there is a 'pigment darkening process' which occurs on re-exposure and takes place more rapidly than primary melanogenesis. It appears to be due to the retention of unpigmented precursors of melanin from previous tannings and is responsive to longer ultraviolet wavelengths than the full reaction. Whether there is equivalent facilitation of thermal and hypoxic responses appears to have received little attention from environmental physiologists.

The components of individual adaptability evidenced in climatic effects can be identified to some extent at least in responses to the other two major environmental variables affecting human beings: infectious disease (reviewed in Playfair 1982; Langman 1989) and nutrition (reviewed in Blaxter and Waterlow 1985), although totally different physiological systems are involved.

Infection

An almost invariant response to infection is fever or persistent raised body temperature. The functional significance of this is still somewhat equivocal but, notwithstanding the subjective malaise it causes, it is best regarded as a desirable response, i.e. other things being equal individuals who develop fever will deal with infection better than those who don't. In this regard it is worth noting that pyrogenic substances are not only released by pathogens but also by a host's own cells, especially polymorphs. Generating fevers has been used as a treatment for certain diseases, such as tertiary syphilis. The fever response can therefore be considered as an example of 'alarm' adaptability, since it occurs quite rapidly on infection. Other features of inflammation can probably be regarded in a similar light.

The main mechanism, however, in dealing with infection is, of course, the immune response with the production of antibody which destroys the infective agent. Even the production of humoral antibody involves complex cellular processes and the multiplication of clones of β lymphocytes.

It typically takes time to develop, and in this sense at least can be thought of as analogous to 'acclimatization'. And here there is strong evidence for cellular memory. Second-set reactions (i.e. caused by repeat stimulation with the same antigen) are substantially shorter than primary ones.

The growth component of individual adaptability appears to have little representation as such in resistance to infectious disease, but at least for endemic diseases developmental factors can be important. Adults, for example, tend to be much more resistant than children. This appears to be largely due to the high levels of immunoglobin they permanently possess as a consequence of repeat stimulation of the immune system and the considerable 'longevity' of many antibodies.

Whilst growth adaptability is little evidenced, genetic adaptation is widespread. The various haemoglobinopathies, which are found in regions of *falciparum* malaria and afford protection from the disease, especially in young children with low immune antibodies, afford an excellent example, but there are a number of others (see Chapter 1).

Nutrition

The concepts of adaptation and adaptability have only recently been taken up by most nutritionists. Customarily human beings have been regarded as instantly adaptable over a whole range of dietaries except for some universal nutritional requirements which always have to be met. It was of course recognized that sharp dietary changes could cause alimentary upset and that insufficient food caused growth retardation and, in extremis, marked physiological changes. Such phenomena were not, however, usually considered within any paradigm of adaptation. The situation has been gradually changing (see Waterlow 1990).

Alarm reactions are of little consequence because of the existence of body stores and reserves for many nutrients, but nonetheless some changes can begin to occur quite rapidly. It is now well established that BMR declines under conditions of energy shortage and the onset of this decline is detectable within hours (James *et al.* 1990). Other evidence for a nutritional equivalent to acclimatization has recently been demonstrated in response to Vitamin A deficiency (Villard and Bates 1986), calcium (Norman 1990), and iron (Cook 1990) shortage. There are now strong indications that the protein requirements of peoples may decline under nutritional conditions of protein deficiency.

Attention has also refocused upon the relationship between growth rate and adult size on the one hand and nutritional supply on the other. Environmentally caused reduction of growth rate has usually been regarded as disadvantageous and indicative of poor health. However, it can also be seen as an example of developmental adaptability facilitating survival in an environment of limited nutrients. Slow growth requires less food and tends to produce small sized adults who have quantitatively lower absolute requirements. Although size can be related to strength, most human tasks can be performed by the small as well as the large, and some better. There is also no evidence that the environmentally small are at any disadvantage by comparison with the genetically small, though information on this issue is sparse, especially in relation to tasks of endurance.

It is important to emphasize here that one is *not* saying that slow growth and small size is a generally desirable human condition if one concludes that they are adaptations to poor nutrition. All one is saying is that under conditions of poor nutrition those who grow slowly will be more successful (in this context healthy) than those who grow rapidly. Clearly the comparative quality of environments is a quite different factor and adaptations do not make poor environments good ones. There has been much confusion over this point in the nutrition literature.

However one may interpret environmentally caused variations in growth and size, there seems little doubt that nutritional factors, as well as climatic ones, have acted as selective agents producing genetic differences in growth and size between populations. Groups who are genetically small in size predominate in tropical ecosystems and where economic circumstances, such as swidden agriculture often lead to regular 'hungry seasons' and chronic malnutrition. Anthropometric variation has been shown to be associated with reproductive fitness in some ecological circumstances. There are also less contentious examples of genetic adaptation to nutrition, such as the life-long production by intestinal cells of lactase in populations which rely heavily on dairy produce as sources of food (reviewed in Durham 1991).

Stress

The components of individual adaptability so far considered relate quite specifically to particular environmental stimuli: the separate components of climatic, disease, and nutritional variation. It would appear, however, that there is also a non-specific physiological response which acts to reduce the effects of any stress generated by an environment. The essential focus of this stress response is the cortex of the adrenal gland which, under the control of the pituitary and the hypothalamus, secretes adreno-cortical hormones. These hormones have wide-ranging effects on carbohydrate, protein, and lipid metabolism, electrolyte and water metabolism, renal function, maintenance of blood pressure, the central nervous system, inflammatory changes in tissues, and exposure to infection, injury, severe muscular exercise, and extremes of temperature. The important hormones from the stress point of view are cortisol and cortisone, and they are thought to play the key role in the General Adaptation Syndrome as recognized some years ago by Selye (1956). He identified three stages to this syndrome, which occurs under severe environmental challenge. First the alarm or emergency reaction, second the period of resistance, and finally the stage of exhaustion and collapse if the challenge persists. Clearly there is some general similarity between the first and second stages, and the first two identified in specific physiological adaptability. Other stress hormones are also known, of which adrenaline, secreted by the adrenal medulla, is probably the most important. It is known as the 'flight–fight' hormone and interacts with cortisol responses, especially during the alarm phase. These hormones are not only of importance in relation to physical environments but even more so to psycho-social ones. They provide insight into the effects of environmental change on human psychological adaptation. Both rapidly respond to new mental experiences with adrenaline mainly associated with 'arousal' levels, and cortisol with feelings of 'affect'. They should probably be seen as components of behavioural adaptation, even when the environmental stimulus is totally physical (Frankenhaeuser 1982).

Adaptability and selection

In this summary statement of components of physiological adaptation, it has been taken as self-evident that evolution through natural selection would produce the responsible mechanisms. All environments are subject to change so homeostatic mechanisms which 'isolate' essential function from the effects of such change will be favoured. When, however, one examines the actual environmental circumstances that are required, it is not at all obvious how some of the processes have come into being. Take, for example, the adaptability to high altitude. This appears to be a general species characteristic, but most human populations have never experienced high altitude or hypoxic environments at any stage in their evolutionary history! Surely organisms need generational exposure to the environmental conditions to which they can mount adaptable responses. In this case one suspects that at least the 'acclimatizing' mechanisms were evolved for other purposes. Some of the features of high altitude adaptability, particularly cardio-vascular sufficiency and effective lung function, are also associated with sustained physical work and exercise and consequential physical fitness. More importantly, however, are the blood and cellular responses. These appear to be triggered by oxygen shortage at the tissue level. Such shortage will also be experienced in anaemia and after substantial blood loss. Blood loss through injury and wounding must have been a common experience in all populations throughout human evolution. Perhaps we should therefore see the general capacity for high altitude acclimatization as a fortuitous expression of homeostatic mechanisms designed for other environmental demands. This, of course, is a quite separate issue from the one of genetic adaptations found in populations who have long lived at high altitude.

Even the processes of thermal acclimatization cannot be easily explained from an evolutionary standpoint. Diurnal variations are of too short a cycle to be relevant, so unless individuals have regularly migrated to and from very different climatic zones, one would expect heat/cold acclimatizing ability to be the product of marked seasonal thermal changes. However, most of human evolution occurred in tropical savannas, mainly African, and the degree of seasonal temperature variation is not particularly marked. Admittedly heat stress will be affected by humidity, which does vary seasonally, but cold, if it is experienced at all, only occurs at night or for short periods.

One cannot help feeling that perspectives have been influenced by the fact that most environmental physiological research has been undertaken by workers from temperate latitudes who do experience marked seasonal temperature variations and find that they can physiologically adapt when they move to both the tropics and the arctic. Seen from a tropical viewpoint, a state consisting of all the features which characterize heat acclimatization might be the norm. The question then becomes: is there a general cold adaptability common to the species? The evidence for this outside behavioural responses is much more problematic. Certainly human beings could not occupy the cold regions or even the temperate regions of the world without the technology of fire, clothing, and housing. People who have long been exposed to cold appear to have some specific physiological adaptations, such as cold-induced vasodilation, but these are not components of general adaptability and appear to be lacking in tropical groups. Contrariwise, tropical peoples who experience night cold as, for example, Australian Aborigines, may have their own genetic adaptations to deal with this specific stress, such as cold-induced hypothermia. These adaptations have to be consistent with the experience of great day-time heat. It has yet to be tested whether prolonged exposure to the cold generally induces raised BMRs as it does in Europeans.

Such concerns, real as they are, do not invalidate the conclusions that human beings have a high level of physiological adaptability to environmental variation and that such adaptability as a species characteristic is the product of natural selection. The value of 'acclimatizations' is self-evident in environments seasonal for the relevant environmental factor. And this is as true for disease and nutritional seasonality as for climatic change. Then one should not assume little year to year variation. Constancy has probably always been a feature of equatorial forest ecosystems, but elsewhere there can be years of excessive heat and cold, drought and food shortage, and epidemic disease, etc. One also knows of major systematic changes of climate over evolutionary time, with all that that implies about other environmental circumstances. The fossil and archaeological record provides testimony of substantial geographical migrations during the later phases of hominid evolution, and the consequential environmental changes must have been considerable, even if people were undertaking migrations to try and keep their environments constant as ecosystems changed their global distributions. In all such circumstances individual adaptability is at a premium.

In considering any aspect of physiological adaptation it is easy to see the advantage of mechanisms which ensure survival at times of environmental crisis: thermoregulatory systems which can obviate excessive heat or cold, metabolic systems which reduce the impact of energy and nutrient shortage in famine and the hungry season, and immunological systems which can cope with epidemic diseases. There are, however, many ways in which the action of natural selection can be much more subtle. Here it is important to remember that selection operates over the whole life-history of an individual from conception to death, and even post-reproductive differentials can be genetically important through kin-selection. Different stages of the life-cycle are varyingly vulnerable with infants and children often at greatest risk. Strangely children have been largely overlooked by environmental physiologists.

It is also important to remember that the environments of individuals living in the same place and at the same time are often not at all the same. This is especially true for a social organism like man, where there is usually some social stratification of environments (Harrison 1990). This stratification frequently leads to varying quality of the biological environment, with individuals in the socially 'superior' strata having more resources, better food, and greater access to health care, etc. Clearly any physiological ability which facilitates access to the better environments will be strongly favoured through the greater survival, reproduction, and offspring survival that the better environments are likely to promote. In other words Darwinian fitness will tend to be highest in 'upper' classes, especially in the absence of contraception, and physiological ability can influence the probability of being in those classes. It is likely to do so through its effects on health and well being. Individuals well adapted to the demands of their daily lives, social as well as physical, will have a strong sense of well being which will allow them to undertake the endeavours and work required to remaining in, or being promoted to, the best available environments. Here the survival of themselves and their children will be favoured. Those whose physiology does not promote health and well being will tend to decline into the least favourable environments where their physiological inadequacies will become even more of a liability. A possible example of the connection between physiological fitness and Darwinian fitness concerns sugar cane cutters in Tanzania. Among them work capacity in the heat determines the amount of cane that can be cut each day. Cane cut determines income which is probably the most important factor influencing socio-economic status. This status affects such important demographic events as age of marriage, number of live births and survival rates of offspring. Thus there exists a direct connection between thermal fitness and reproductive fitness. One can think of many other possible sequences like this one.

Fitness

Although we have written largely as though the physiological changes which can be seen on environmental changes represent different settings of homeostatic regulatory mechanisms, it has also been clearly recognized already that environmental change will be associated with changes in fitness. At the physiological level this fitness is equatable with health and well being, i.e. with homeostasis itself.

Some environmental changes will be too great for homeostasis; new environments may contain components which have never been experienced before, either qualitatively or quantitatively, and even where homeostasis appears to be occurring differing environments are likely to be variably optimal. If two environments differ in their optimality then, by definition, they must differ in at least the ease with which homeostasis can be achieved or maintained. In all these situations there will inevitably be a change in the fitness in the individuals experiencing them, and this is likely to be reflected in the state of a number of characters of the individuals concerned. Character changes, therefore, may not be reflecting changes in homeostatic settings but changes in fitness.

In many cases it is self-evident when a character change represents a fitness change. All those features of biochemistry, physiology, or morphology which are medically defined as 'morbid' are by definition indicative of loss of physiological fitness—and are also likely to be indicative of loss of Darwinian fitness. More generally in any physiological system where the role of various physiological mechanisms is to maintain some physiological state stable, loss of that stability represents a loss of fitness, i.e. the homeostatic systems were inadequate for the circumstances. The human body tends to 'work best' within relatively narrow ranges of body temperature, blood H-ion concentration, osmotic pressure and sugar concentration, and various other levels of cellular and body fluid constituents. Marked changes in these characters tend to reflect failure of homeostasis, and therefore decline in physiological fitness.

It is not always obvious, however, whether a character change represents a fitness change or an adaptive response. This is particularly the case in developmental processes and especially growth itself. Hence the controversy about the meaning of the relationship between growth and nutrition. Growth is strongly target-seeking under elaborate homeorhetic control. This is particularly evidenced in the phenomenon of 'catch-up' growth. A child whose growth is retarded by ill-health will grow especially rapidly once it recovers so as to make-up the stature it failed to achieve during the insult (Tanner 1986). Full catch-up, however, only occurs after rather short insult. The longer the insult the less the capacity for full catch-up and, of course, if the insult persists throughout growth the child will be permanently stunted. This is the basis for concluding that slow growth (once genetic factors have, if possible, been allowed for) is a fitness indicator of adverse environments.

However, there is an alternative view which sees varying growth as an adaptable mechanism. In this view, every genetic constitution is endowed with a range of responses to environments which an organism is likely to experience (the norm of reactions as defined by Dobzhansky (1962)). Thus there are a multiplicity of possible growth patterns for an organism, and the one followed is the one most appropriate to the environmental circumstances. There are certainly examples where growth behaves in an adaptable way. The tail of the mouse is an important thermoregulatory structure, and in high temperature environments grows much more rapidly and longer than in cold ones. In humans the thoracic growth and development which occurs when indigenous lowlanders are brought up in hypoxic high altitude conditions should likewise be seen as adaptable.

The problem of distinguishing fitness changes from adaptable changes is not limited to growth. Humans lose body weight at extremes of altitude, but it is far from clear how this should be interpreted. Further, it must be recognized that in extreme environmental circumstances the homeostatic/adaptable mechanisms will begin to break down and thus in effect become fitness indicators themselves!

It is also being increasingly recognized that not only is long term development a component of adaptability, it is also a component of fitness. A number of cases are now known where environmental experiences in childhood are responsible for later morbidity in adults though apparently having no ill effects at the time (Ciba 1991), and many constitutional diseases take a long time to produce recognized morbidity. During such latent periods the nature of the changing physiological processes may be hard to interpret.

Fitness and variability

It has been suggested that comparison of levels of character variability can help to resolve the problem (Harrison 1966). Obviously fitness characters will show comparatively small components of environmental variance between individuals when environmental circumstances are favourable. Under such conditions homeostasis is, by definition, efficient so the effects of inevitable environmental heterogeneity and perturbations will be minimal. By contrast in adverse circumstances homeostasis is under strain and even quite small environmental differences may be expected to have large, and perhaps permanent, biological effects on fitness. Fitness characters will therefore tend to show increased variability in populations.

By contrast one might expect the variability of homeostatic mechanisms to behave, if anything, in a completely opposite way. In favourable environments, where homeostasis is easy, there are many different settings of the underlying mechanisms compatible with high fitness. Any one mechanism would therefore tend to show high environmentally caused variability. Conversely, adversity will demand that all relevant mechanisms be operating to full effect; in which case they will each show lowered variability.

There are difficulties in applying these expectations to real situations. In principle, one is the comparative size of genetic components of variability. These could clearly differ from one population to another. However, it can be argued that differences in the magnitude of character variability in different populations arising from genetic causes are likely to be very small compared with variability from environmental causes in characters of low heritability, such as adaptable mechanisms and fitness traits are bound to have (Harrison 1987). The absolute genetic component can probably be taken as very nearly a constant.

More troublesome are likely to be differences in the amount of environmental heterogeneity populations are exposed to. Clearly environments which are very heterogeneous are going to cause greater variability than ones which are fairly homogenous at the same average level of favourability. However, one might expect strong interactions between heterogeneity and favourability, with varying levels of environmental heterogeneity having little or no effect in broadly optimal circumstances on, at least, the magnitude of variability of fitness indicators.

By contrast in adverse environments even small differences in levels of heterogeneity are likely to have notable effects, since with homeostatic mechanisms under strain even minor environmental fluctuations will influence fitness. There is empirical experimental evidence in animals in support of these propositions (Harrison 1963).

In the few situations which have so far been examined, it would appear that human growth, specifically in stature, is more variable in overall adverse environments than in ones that are regarded as desirable (Harrison and Schmitt 1989). This is especially evidenced in variability of short term increment where differences between populations can be considerable (Harrison *et al.* 1990; Brush *et al.* 1992). It suggests that growth should be primarily seen as a fitness indicator rather than as an adaptable mechanism.

Many features of physiological adaptation can be summarized by considering the fitness changes and the population variability changes of physiological characteristics when individuals face a marked environmental change. Typically the environment in which a population evolved, its habitual environment, will be optimal for it. In the human case such a generalization needs qualification since, through technology, humans can remove from all environments many if not all of the natural hazards which affect optimality. Nevertheless, the fact that there are genetic adaptations to particular environmental circumstances implies that different human groups have different requirements for optimality. And more significantly some environments are generally more suitable for human existence than others. We can therefore reasonably consider what happens on transfer from favourable to adverse conditions. This is schematically shown in Fig. 2.1. By definition there must be a loss of fitness on such transfer. Just how rapidly this loss occurs would appear to be dependent upon the nature and form of the insults. In the case of thermal, and especially hypoxic stress, loss occurs very rapidly and lowest fitness is experienced within days if not hours or even minutes of the transfer. By contrast, fitness will be lost slowly on moving into an environment of nutritional shortage, because of the existence of body stores and reserves of nutrients. In this case one can speak of a 'carry-over' of fitness with good nutritional health in one environment protecting for a time against the insults of the new. There is little or no such carry-over in climatic ecology, even good general physical fitness being of minor advantage. The situation with infectious disease seems to lie somewhat intermediate, with the main determinant of fitness change dependent upon the epidemiological factors which influence the likelihood of being infected with a new pathogen and of that pathogen multiplying.

In all these circumstances, however, one would expect the fitness of survivors to rise after some time. This will be due to the development of

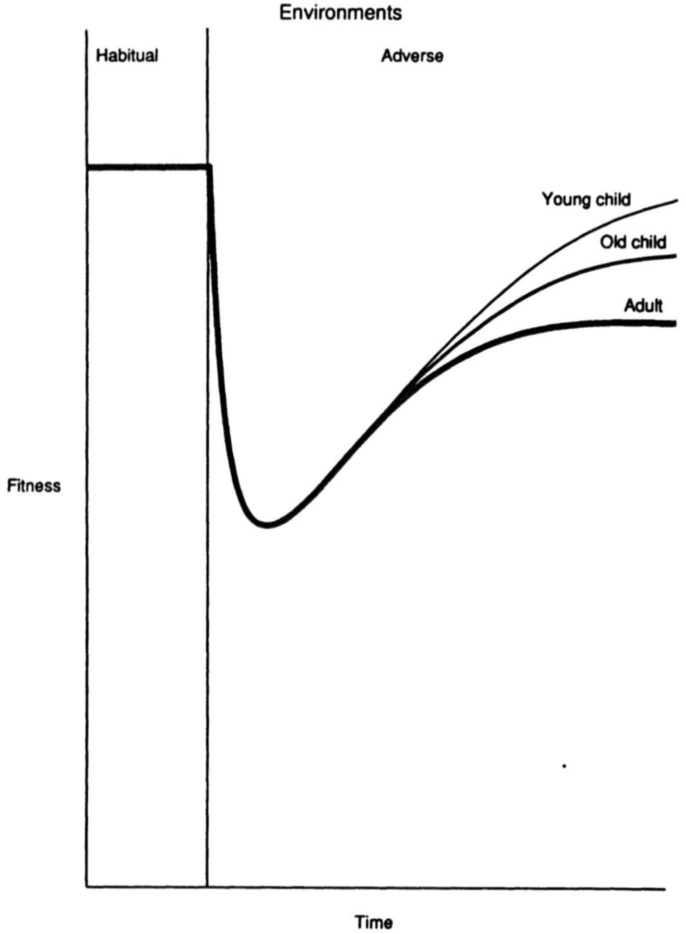

Fig. 2.1 Comparative changes in the overall fitness of individuals on movement from habitual to adverse environments.

the 'acclimatization' kind of adaptation. For climatic change itself a large proportion of the ultimate level of full acclimatization is achieved quite quickly, within 14–21 days in heat and hypoxic stress. In the case of nutrition and disease processes, for various reasons environmental as well as intrinsic, it appears to take longer. Eventually a new equilibrium of fitness is established but in most cases this will be at a lower level than in the original environment. Because they are capable of undertaking developmental adaptability the rise in fitness in children will tend to progress further than in adults, and the extent of this extra rise should be age-related: the younger the child on transfer the greater the additional rise. Of course, if there is a return to the original environment the developmental adaptations to the adverse conditions may turn out to be a handicap, though probably of no great significance in overall good circumstances.

The predicted behaviour of between individual variability in a population which experiences the same kind of environmental change is shown in Fig. 2.2. Confidence is greatest for the 'fitness character' predictions—an initial rise in variability with a gradual subsequent fall

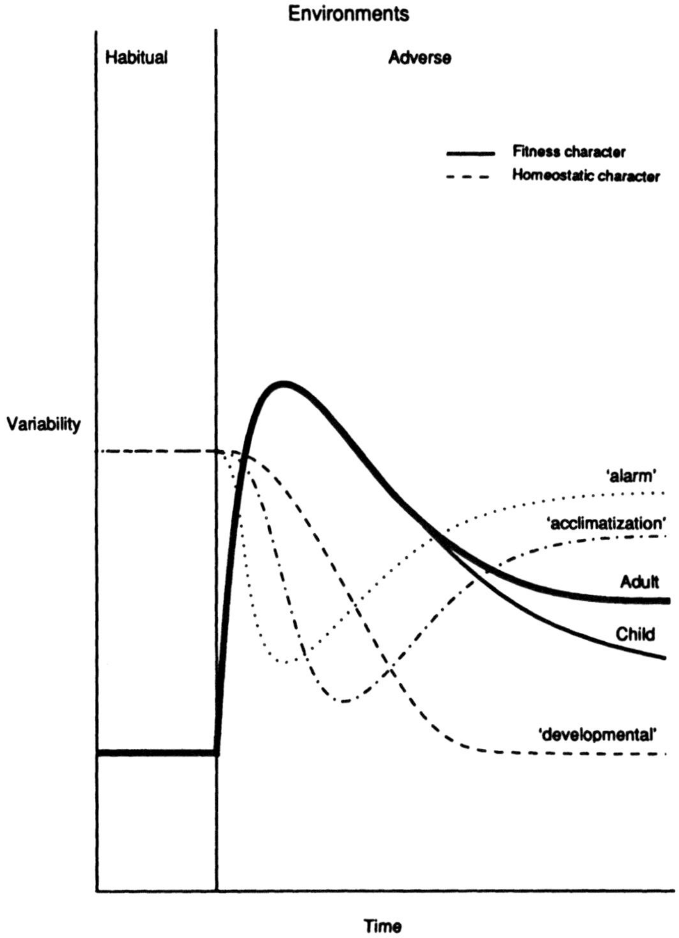

Fig. 2.2 Comparative between-individual variability in fitness and homeostatic characters on movement from habitual to adverse environments.

as adaptation progressively occurs with increasingly efficient homeostasis. Because of developmental adaptability it seems likely that the variability in groups who were first exposed to the adverse conditions as children will be less than in adult 'migrants', but not withstanding their more successful adaptation, variability at equilibrium will still be greater than in the original favourable environment.

The comparative variability of the homeostatic mechanisms themselves is more speculative. If the arguments presented earlier are right one might expect a sequence of reductions in variability affecting first alarm mechanisms, then acclimatization ones, and finally developmental ones. The latter should remain very homogenous and at variability levels well below those found under favourable conditions (after due allowance for any effects of mean character value), but there might well be some rise again in alarm and acclimatization character variability as these mechanisms are progressively replaced, especially the former. Unfortunately, there seems no information against which these predictions can be tested.

Conclusions

The concept of adaptation and adaptability has provided an important framework in human physiology, it having long been recognized that human beings have remarkable abilities to maintain internal environments constant in a great diversity of external environments. Physiologists in fact have been much more able to demonstrate unequivocal adaptability than biologists have genetic and evolutionary adaptation. This of course is because adaptability is open to rigorous, controlled, and reductionist experimentation (which incidentally has led to a diverse vocabulary of physiological classes of adaptation not mentioned here). However, physiologists have been almost exclusively concerned with the elucidation of mechanisms of adaptability. These are frequently very complex in detail, and always much more complex than as described in this broad ranging review. What physiologists have tended to ignore is the biological *effects* of the mechanisms they have discovered, and particularly the variability in these effects. In most physiological experimentation variability is a nuisance—a noise that acts to conceal the truth. But to an evolutionary biologist, and as I hope I have shown here, to a population ecologist, variability is the truth to be explained. When it has genetic components it is the raw material of evolution, and when it has environmental components it can tell us much about the nature of the relationships between human populations and their immediate environment.

References

Baker, P.T. (ed.) (1978). *The biology of high altitude peoples.* Cambridge University Press.

Baker, P.T. (1988). Human adaptability. In *Human biology* (ed. G.A. Harrison, J.M. Tanner, D.R. Pilbeam, and P.T. Baker). Oxford University Press.

Blaxter, K. and Waterlow, J.C. (eds) (1985). *Nutritional adaptation in man.* John Libbey, London.

Brush, G., Harrison, G.A., Baber, F.M., and Zumrawi, F.Y. (1992). Comparative variability and interval correlation in linear growth in Hong Kong and Sudanese infants. *Am. J. Hum. Biol.*, **4**, 291–9.

Cook, J.D. (1990). Adaptation in iron metabolism. *Am. J. Clin. Nutr.*, **51**, 301–8.

Ciba Foundation (1991). *Childhood environment and adult disease*, Symposium No. 156. Wiley, London.

Dobzhansky, T. (1962). *Mankind evolving.* Yale University Press, New Haven.

Durham, W.H. (1991). *Coevolution.* Stanford University Press.

Frankenhaeuser, M. (1982). Challenge–control interaction as reflected in sympathetic-adrenal and pituitary–adrenal activity. *Scand. J. Psych., Suppl. 1*, 158–64.

Harrison, G.A. (1963). Temperature adaptation as evidenced by growth of mice. *Federation Proc.*, **22**, 691–8.

Harrison, G.A. (1966). Human adaptability with reference to I.B.P. proposals for high altitude research. In *The biology of human adaptability* (ed. P.T. Baker and J.S. Weiner). Oxford University Press.

Harrison, G.A. (1987). *Adaptability, fitness and health.* The First Paul T. Baker Lecture, Pennsylvania State University, pp. 5–18. Pennsylvania State University Press.

Harrison, G.A. (1990). Lifestyles, health, wellbeing and natural selection. In *Primate life history and evolution* (ed. C.J. DeRousseau). Wiley-Liss, New York.

Harrison, G.A. and Schmitt, L. (1989). Variability in stature growth. *Ann. Hum. Biol.*, **16**, 45–51.

Harrison, G.A., Brush, G., Almedom, A., and Jewell, T. (1990). Short-term variation in stature growth in Ethiopian and English children. *Ann. Hum. Biol.*, **17**, 407–16.

James, W.P.T., McNeill, G., and Ralph, A. (1990). Metabolism and nutritional adaptation to altered intakes of energy substrates. *Am. J. Clin. Nutr.*, **51**, 264–9.

Langman, R.E. (1989). *The immune system. Evolutionary principles guide our understanding of this complex biological defense system.* Academic Press, London.

Norman, A.W. (1990). Intestinal calcium absorption: a Vitamin D-hormone mediated adaptive response. *Am. J. Clin. Nutr.*, **51**, 290–300.

Playfair, J.H.L. (1982). *Immunology at a glance.* Blackwell Scientific Publications, Oxford.

Robins, A.H. (1991). *Biological perspectives on human pigmentation.* Cambridge University Press.

Selye, H. (1956). *The stress of life*. McGraw-Hill, New York.
Tanner, J.M. (1986). Growth as a target seeking function: Catch-up and catch-down growth. In *Human growth: A comprehensive treatise* (ed. F. Falkner and J.M. Tanner). Plenum, New York.
Villard, L.F. and Bates, C.J. (1986). Carotine deoxygenase activity in rat intestine: effects of increased activity of carotine deoxygenase in Vitamin A deficiency. *Br. J. Nutr.*, 56, 115–22.
Waterlow, J.C. (1990). Mechanisms of adaptation to low energy intakes. In *Diet and disease* (ed. G.A. Harrison and J.C. Waterlow). Cambridge University Press.

3
BEHAVIOURAL ADAPTATION

Robin I.M. Dunbar

Introduction

Ever since Durkheim's seminal work at the turn of the century (e.g. Durkheim 1915), social scientists have tended to insist that human behaviour (and human culture, in particular) is not adaptive and cannot be amenable to analysis from a biological perspective. Human social behaviour, so Durkheim argued, can only be understood in terms of the social context within which it is performed. From this derives the idea, widespread within the social sciences, that human behavioural and cultural norms are simply a reflection of the societies of which they are a part: it is the society (or culture) itself that determines what the individual believes and does.

This view is in marked contrast to that of evolutionary biologists. Working within a Darwinian paradigm, biologists insist that an individual's behaviour is designed so as to maximize genetic fitness (which for most practical purposes can be—rather loosely—interpreted in terms of that individual's ability to leave descendants). Rather than society moulding the individual in its own shape, biologists have tended to insist that the individual (or at least the collective of individuals who live together) mould their society and its beliefs to suit their own needs.

(In fact, of course, from a strictly biological point of view, an individual's choice of action is always constrained by what the other members of its group, or even community, are trying to do. Consequently, an individual's behaviour genuinely *is* determined, in part at least, by the nature of the society it lives in. Viewed in one light, this can be interpreted as more than just a nod in Durkheim's direction on the part of biologists; nonetheless, the Durkheimian view is in conflict with the biologists at the deeper level to which biologists wish to penetrate in their analyses of behaviour. The difference seems to be largely one of time frame: from the proximate perspective of Durkheim, an individual's behaviour really is determined by the culture he or she is born into, but, in the longer perspective of evolutionary time, individuals do determine their cultures.)

The received view in the social sciences implies that human behaviour is not (and indeed never can be) adaptive in any biological sense. Unfortunately, this view seems to be based on a rather narrow conception of what constitutes an *adaptation* in the sense that biologists use this term: it seems to assume that it refers only to adaptation to the exigencies of the natural world. I shall try to show that adaptation has a much wider connotation in evolutionary biology than this view presupposes and that, seen in this wider context, at least some patterns of human behaviour are adaptive. In addition, however, I shall also try to show that human behaviour can at least sometimes be adaptive in the narrower ecological sense presupposed by the received Durkheimian view. I shall argue that the failure to recognize that behaviour (or culture) is genuinely adaptive is in part at least a consequence of the fact that our understanding of biology has often been too superficial.

This does not, of course, mean that *all* human behaviour is *ipso facto* adaptive. Indeed, there is no biological reason why it should be: evolutionary biology merely claims that organisms' behaviour will, if left to itself, tend towards increased adaptiveness. Degrees of adaptiveness are inevitable since evolution is a dynamic process and the biological world thus undergoes constant change: that, after all, is why species eventually go extinct! In addition, however, there is a branch of evolutionary theory that concerns itself with the evolutionary behaviour of traits under conditions of neutral selection when there is no direct selection pressure that leads to adaptation (see Maynard Smith 1989). For present purposes, however, I merely note this fact.

The problem of measuring adaptation

When we speak of adaptation in an evolutionary context, we mean to suggest that a particular trait confers a selective advantage—in other words, that a given trait allows its possessor to contribute more copies of a given gene to the next generation than any alternative trait of the same kind. This is the benchmark against which evolutionary biologists judge a trait's adaptiveness.

However, it is important to remember that this is not the end of the matter as far as evolutionary explanations go. To assert that a trait contributes to the propagation of genes in future generations is to say no more than that those genes that survive are the ones that survive best, the classic 'evolutionary tautology' (see Dunbar 1982, and references therein). Darwinian explanations should offer us more than this: they should tell us *why* some genes are propagated more effectively than others. This part of the classic Darwinian formula is concerned with adaptation. This usually requires us to say something about the way in which a given trait allows its possessor to solve one or more of the problems of survival or reproduction well enough to contribute genes to the next generation.

Conventionally, evolutionary biologists have accepted evidence for either of these two components as being evidence for evolutionary adaptation. One reason for this is that Darwinian explanations are explanations about the genetic *consequences* of acting in a particular way. It necessarily follows that if a trait allows an individual to solve a problem of survival or reproduction more successfully than do any alternative traits, then it must contribute more copies of any relevant gene to the next generation. Conversely, if individuals who possess a given trait contribute more copies of a gene to the next generation than those possessing any alternative traits, then it will usually (though not always!) be the case that such a trait must have allowed them to solve problems of survival and/or reproduction more effectively.

In asking whether human behaviour is adaptive, then, our concern is principally with the question of whether behaviour is functionally related to some problem of survival or reproduction. We can demonstrate adaptiveness either by showing that the actor has higher genetic fitness than individuals that do not behave in that way, or by showing that the behaviour in question allows an individual to solve a particular problem of survival or reproduction efficiently (in an engineering design sense). We might show, for example, that, given (i) the nutritional value of different types of prey, (ii) the frequencies with which they are encountered by a foraging animal, and (iii) the ease with which it can catch and eat them, an individual would gain more calories per unit time spent hunting when it hunts large slow moving prey than when it hunts small fleet animals; if individuals do in fact turn out to prefer large sluggish prey more often when they have a choice between them (or prefer to continue searching for such prey for longer before switching to a small fleet one), then we can infer that the hunting rule adopted by the individual is adapted to its circumstances.

Having said this, however, two caveats are in order. Both are a consequence of the fact that the real world is invariably more complex than our simple theories like to pretend. One concerns the fact that contributing genes to the next generation is not simply a matter of solving the problem of how to mate most often. The other arises from the fact that animals can trade short-term losses against long-term gains to give rise to alternative strategies that are equally good (but radically different) solutions to the same biological problem.

The first of these problems is a direct consequence of the systemic nature of biological organisms. An organism is, in some very real sense, a coadapted system: it has been designed to solve a number of (often mutually conflicting) problems simultaneously. This is because genetic fitness (the replication of genes into future generations) is nothing more than a consequence of the business of reproduction. But reproduction itself is a complex process, especially in advanced organisms like mammals: it requires the animal concerned to survive successfully, to find and court mates successfully, and then to produce healthy offspring from those matings that will themselves live long enough to reproduce in their turn.

Now, in an ideal world, the perfect solution to the problem of reproduction is clearly to mate (or give birth) as often as possible. But that is only possible in an ideal world: real animals have to persuade reluctant and choosy members of the opposite sex to mate with them in the first place; they also have to eat enough to ensure that they can continue to engage in mating sufficiently often to out do their rivals. If obsessive interest in mating means that you are caught unawares by a predator or die of starvation through failing to eat, then you will inevitably produce fewer offspring (and hence contribute fewer copies of your genes to the next generation) than another individual who behaves more circumspectly. In biology, an optimal solution to a particular problem need not always be a perfect solution because of the need to achieve a working compromise for the system as a whole. This does not mean that you *cannot* have a perfect behavioural adaptation; it is simply a caution against assuming that an apparently poor fit between a problem and the behavioural solution adopted by an animal implies that behaviour is non-adaptive. In many cases, the match between problem and solution will be a good one, but if the problems created for the animal along other dimensions of the system are particularly intrusive, then the fit may be much poorer. This is very inconvenient, of course, but it is something we just have to learn to live with in the real world.

This gives rise to the second problem, namely the fact that individuals can trade costs and benefits on one dimension of the system with those on another. This is likely to be especially common at the behavioural level because species which possess behavioural flexibility will often find alternative ways of solving the same problem. Paradoxically, perhaps, such alternative solutions may in fact be a consequence of the success of a given behavioural strategy. Being aggressive, for example, may prove to be a highly successful mating strategy because it allows the male to keep rivals away from the females. However, such situations lead naturally to evolutionary 'arms races': only the most aggressive sons of an aggressive male succeed in mating in the next generation, and only the most aggressive sons of these aggressive sons do so in the third generation. As a result males become increasingly aggressive with each passing generation. As they do so, they incur increasing costs: the wear and tear of high testosterone and stress levels makes them susceptible to diseases, and the high rates of fighting inevitably result in increased risks of injury and death. In addition, males that are too aggressive are likely to frighten off the females they are trying to court. Once these costs begin to rise steeply, it will start to pay males to behave non-aggressively. Non-aggressive males may mate less often each year, but they will survive for much longer. Providing (1) they can obtain at least a few matings each breeding season and (2) they can achieve proportionately longer lifespans than aggressive males, they will be able to contribute as many genes to the next generation as aggressive males. This can, in principle, result in a balanced behavioural polymorphism that is an evolutionarily stable strategy (or ESS). In effect, both males are trading short and long term benefits against each other, but doing so in complementary ways.

The problem that this creates for the biologist is knowing enough about the system as a whole to be able to recognize the difference between a genuinely maladaptive strategy and one that merely *appears* to be maladaptive. We observe the success of the aggressive males and correctly infer that it is an adaptive solution to the problem of mating. However, we also find an unexpectedly large proportion of males pursuing what appears to be the decidedly sub-optimal (and hence maladaptive) strategy of sitting on the sidelines and not competing for access to the females. We are tempted to infer from the rather poor performance of the non-aggressive males in each breeding season that such behaviour is an evolutionary failure: the male's genes are heading for rapid extinction. We can only recognize the true situation when we can take a long enough time perspective and see the whole process from birth to death. Evolution is a function of lifetime reproductive performance not the success of the moment.

An additional layer of complexity is added by the fact that evolutionary trade-offs of this kind may actually take place across the generations. In a pair of mathematically rather formidable papers, Hamilton (1964) showed that a gene's fitness (the success with which it is replicated in future generations) could be maximized either by the individual bearing it reproducing itself *or* by the individual helping a relative that shared the same gene to reproduce more effectively than it would otherwise have done (or some combination of the two). Hamilton termed this combination of two fitness components the gene's *inclusive fitness*.

One consequence of this reinterpretation of the way in which evolutionary processes work is that genes can evolve under conventional Darwinian processes even though they result in the death of the individual that possesses them. Altruistic behaviour (in which an individual sacrifices itself for the benefit of others) could in fact be a stable evolutionary strategy even though at first sight it looks, in Darwinian terms, to be quite maladaptive. The conditions under which it is possible for a gene to evolve by this process are known very precisely (usually referred to as *Hamilton's Rule*). The general theory based on this idea is generally known as the *Theory of Kin Selection*. I do not wish to devote any time to a consideration of kin selection here; I raise the issue simply to point to yet another kind of perfectly conventional biological mechanism that can produce behaviour that seems, on the face of it, to be maladaptive. Anyone wishing to pursue this issue in more detail can find suitable treatments of it in Dawkins (1976), Barash (1982), and Daly and Wilson (1983).

One final point should be noted: biological explanations about fitness and adaptiveness do not necessarily presuppose that behaviour is genetically determined. They require only *heritability*, a term that has now generally come to be defined in terms of 'fidelity of copying' (i.e. the accuracy with which offspring resemble their parents). The mechanism that facilitates such copying is entirely open: it might be DNA (and thus involve conventional biological reproduction and inheritance) or it might involve learning. The mechanism by which a trait is transmitted from one generation to the next (i.e. from parents to offspring) does not affect whether or not evolution (and hence adaptation) occurs, though it may influence the *rate* at which evolution can occur. I am not particularly concerned here to paddle in the murky waters of the nature/nurture controversy (most of which is rather unproductive). I simply want to point out that the discussion that follows makes no assumptions at all about the *mode of inheritance*. Indeed, I shall in the final section discuss one example for which any kind of conventional genetic basis can almost certainly be ruled out.

A peculiarly human problem

Humans are among the most long-lived of organisms, being exceeded in lifespan by only a few relatively unusual species like oak trees and giant tortoises. Since the observer is also human, it becomes difficult to assess the lifetime reproductive performance of one's subjects. Indeed, with any species that has a potential longevity in the order of 80–100 years there are serious problems: few sources of research grants are willing to commit themselves over such an extended period of time. This means that while it is easy to see how human behaviour is related to the immediate problem it is designed to solve, we can seldom see the likely consequences in terms of lifetime reproductive output.

Humans, however, have one redeeming feature: they sometimes keep records of their births and deaths. This means that, for some populations at least, it is possible to determine the number of descendants that a given individual leaves over periods as long as two centuries or so (six or seven generations). This is as close to true measures of genetic fitness as it is possible to get (see Prout 1969). To be sure, these records are not always perfectly reliable: people have not always been accurate or consistent in their registrations, and they have sometimes deliberately sought to cover up socially inconvenient events (such as the birth of illegimate children). There are, in addition, problems associated with the reliability of paternity: while female mammals can generally be certain that the babies born to them are in fact their own genetic offspring, male mammals can never be so sure. Nonetheless, within the level of potential error introduced by these kinds of biases, historical registers of birth and death allow us to determine quite reasonable estimates of an individual's fitness. The catch, of course, is that we cannot relate these measures of fitness to individual behavioural strategies because we do not have a sufficiently detailed record of people's behaviour so far back in time. The best we can do is compare the fitness of rather general classes of individuals rather than of behavioural strategies.

This is very unfortunate: it means that we cannot marry up the two components of the Darwinian formula and show that adaptiveness in the engineering sense actually does result in higher genetic fitness. In this sense, it seems to constitute a genuine biological 'Uncertainty Principle'. As with Heisenberg's Uncertainty Principle in quantum physics (which says that you cannot simultaneously measure both the position and the speed of a particle), it seems that you cannot ever know both the adaptiveness and the fitness consequences of a given behavioural strategy for a long-lived species.

Nonetheless, we need not be overly pessimistic. We can be fairly certain that behaviour that is adaptive (in the engineering sense) will also confer higher fitness. The real problem is that we can never be sure that behaviour that is poorly adapted has low genetic fitness. Some of these cases will surely be genuine cases of maladaptation (caught, as it were, on their way to extinction), but others will be optimal compromises or strategies in alternative strategy sets.

In effect, this means that we are ill-advised to attempt to develop evolutionary theory using humans as our subject species. Humans are not, of course, the only species for which this will be true. To a greater or lesser extent, we face the same problems with any long-lived species. However, since evolutionary theory is necessarily assumed to hold universally, we can use the theory developed using demographically more tractable species to explore and explain human behaviour.

Humans do, however, have one other characteristic which creates difficulties for evolutionary analyses of this kind: this is culture. Culture is interesting from an evolutionary viewpoint because it involves the transmission of learned rules of thumb about how one should behave. Irrespective of the inertial effects of the cultural transmission mechanisms (i.e. the fact that we seem to be predisposed to learn cultural rules from those around us at such an early age that these rules become quickly and very firmly engrained), cultural transmission has at least two advantages over genetic transmission for long-lived species. First, it allows much greater flexibility of response to changes in the environment that occur over time periods shorter than the natural generation length. Long-lived species would soon become extinct without the ability to learn new responses quickly. Secondly, the cultural inheritance of prepackaged rules of behaviour incurs considerable time savings for the majority of those who acquire the rule. A new behavioural response can always be learned by each individual on a trial-and-error basis, but the same mistakes will be repeated by every individual. There are obvious advantages to be gained if particularly effective rules discovered by one individual can be passed on to others. Cultural transmission processes make that possible.

However, it soon became apparent to biologists that cultural transmission can take at least two different forms. In one, culture is essentially parasitic on conventional genes because it ultimately depends on the bodies (or brains) created by genes for its survival. In some cases, culture can function as a means whereby genes further their own survival. In other cases, genes and culture may enter into a symbiotic relationship similar to that which often evolves between parasites and their hosts. These mechanisms give rise to the kinds of gene–culture co-evolution models discussed by Lumsden and Wilson (1980). They assume that culture is parasitic on genes and thus ultimately always constrained by what is in the gene's interests. Culture is, in some sense at least, the genes' way of increasing their own rate of replication.

However, it is also in principle possible for cultural elements to acquire a life of their own that is quite independent of the genes whose bodies (or brains) they occupy. Dawkins (1976) pointed out that any self-replicating entity is, by definition, a Darwinian entity that will behave under the impact of natural selection in just the same way as more conventional genes do. Dawkins gave these entities the name *meme*. Memes are things like rules of behaviour or items of information that can be learned and passed on. Dawkins recognized that, in principle at least, such non-biological replicators could live in their own Darwinian universe that is quite independent of that occupied by genes. In effect, memes have memetic fitness that they seek to maximize through replication just as genes have genetic fitness. One implication of this is that the Darwinian interests of memes and genes might conflict (as, for example, if a meme for suicide appeared). Theoretical analyses of cultural transmission developed by Cavalli-Sforza and Feldman (1981) and Boyd and Richerson (1985) have shown that, under certain circumstances, the faster-replicating memes can in fact drive conventional genes to extinction, yet still survive themselves. This view in which genes and memes occupy separate universes constitutes what is usually known as *dual inheritance theory*.

The implications of this for evolutionary analyses of human behaviour will be glaringly obvious: some behaviour patterns may adversely affect the survival and reproductive opportunities of the individuals who inherit them and yet evolve quite naturally under Darwinian selection processes. If we measure the conventional genetic fitness of the individuals adopting them, we will discover that the behavioural rules concerned appear to have negative fitness. However, an alternative explanation that we always have to consider is the possibility that we have been measuring the wrong kind of fitness. Under the dual inheritance theory, we need to measure *memetic fitness* (i.e. the rate at which the memes replicate themselves) not genetic fitness.

Because we do not always know which universe we are operating in in any given case, the most appropriate rule of thumb should always be to consider conventional genetic fitness first. Only if we find a poor match between behaviour and appropriate measures of fitness should we consider the situation in terms of memetic fitness.

We have, of course, to be careful not to overlook the more complex features of genetic fitness discussed in the previous section: it would be very tempting to assume that a lack of fit between behaviour and the problem the organism appears to be solving is a consequence of dual inheritance processes when in fact it reflects the existence of an alternative strategy set or a more conventional compromise solution to a set of conflicting biological problems (i.e. a trade-off between the demands of two different components of the system). We can obviate this problem only by carefully testing the more obvious possibilities first and then applying a kind of Sherlock Holmes procedure: if we have excluded all the obvious possibilities and the only explanation left is the unlikely one of memetic evolution, then that must be the right explanation, however unlikely it might seem.

My point here is simply a cautionary one. Biological systems are enormously complex, often considerably more complex than the much vaunted complexity of human culture. It is only too easy to approach the real world in a naively simpleminded way, and we do so at our peril: it leads inevitably to the 'garbage in, garbage out' mode of science. Evolutionary biologists have learned the hard way not to underestimate the complexity of the real world.

Is behaviour adaptive?

Having cleared the theoretical ground, we can now turn to the rather more interesting task of asking whether human behaviour is ever adaptive. In order to reduce the complexity to a level where we can actually get our minds around the problem, I shall make two simplifying assumptions in discussing examples of possible behavioural adaptation: I shall completely ignore the influence of both kin selection and dual inheritance processes. In many of the cases where these processes are likely to be intrusive factors we do not yet know enough about the details of the social and biological system to be able to undertake a full analysis at this time. Instead, I shall simply try to show that at least some behaviour patterns exhibited by humans do indeed appear to be adapted to solving particular problems of survival or reproduction.

I shall consider three general kinds of problem that behavioural adaptation might be relevant to. These reflect different components of the biological system within which humans, like all organisms, are embedded. These are adaptations to ecological conditions, mating strategies, and parental investment strategies. Finally, however, I shall conclude with an afterthought: an example of what seems to be a genuinely uncontroversial case of a cultural meme evolving.

Adaptations to ecological conditions

The Maasai are a pastoralist people occupying a range of habitats in central East Africa. As with all pastoralists, their cattle are the key to their survival in what are in general marginal rangelands. Not only do the cattle occupy a particularly prominent place in their culture and mythology, but the Maasai are extremely knowledgeable about their animals' requirements and ecological characteristics. One example of the sophistication of their ecological knowledge concerns the relationship between the preferred cattle colours and habitat characteristics. Cattle coat colours vary considerably across the Maasai population, with marked differences between localities in the herdsmen's preferences for dark and light coloured cattle. These differences have usually been attributed to local idiosyncratic preferences on the part of individual herdsmen. However, Finch and Western (1977) examined the distribution of coat colour in Maasai herds across habitat variables and found a very good negative correlation between the percentage of light coloured cattle in the herd and both altitude ($r^2 = 0.80$) and potential evaporation ($r^2 = 0.82$). Essentially, herds had a higher proportion of light coloured animals in warmer habitats.

When Finch and Western examined the physiological characteristics of Maasai cattle, they found that dark coloured animals were less tolerant of high temperatures and required more water (and hence could travel less far each day on a given quantity of water) than light coloured cattle. Light coloured cattle put on weight faster under drought conditions, but dark coloured cattle did better under conditions of higher rainfall. As a result, light coloured cattle generally had higher survival rates under the high ambient temperatures prevalent at lower altitudes close to the equator, whereas dark coloured cattle did better under the cooler conditions prevailing at higher altitudes (where night-time temperatures commonly approach 0 °C).

The Maasai were clearly very much aware of these differences and actively select their cattle to meet these environmental constraints. In habitats with the least heat stress, herds contained about 50 per cent light coloured animals, whereas in habitats with the most heat stress they contained about 90 per cent light animals. Finch and Western comment that the Maasai's expressed preferences for cattle of different colours in different places has sometimes misled anthropologists into thinking that the herdsmen's preferences are quite arbitrary. In fact, their preferences reinforce what nature does to their herds. Indeed, Finch and Western were impressed by the herdsmen's knowledge of their cattle's performance under different environmental conditions: in many cases, the Maasai were able to anticipate the results of Finch and Western's experimental studies or even point them to crucial environmental factors underlying their behavioural decisions. However, Western and his colleagues became aware that the Maasai were often only prepared to discuss these issues once they realized that the scientists could discuss the ecology of the environment in as much depth as the Maasai themselves could.

A second example of ecological adaptation concerns birth spacing in the !Kung San bushmen. The !Kung occupy the marginal habitats on the edge of the Namib desert in southwestern Africa. Although hunting is an important activity, its contribution to the overall diet is very much less than the products of gathering (mainly the work of the women). Gathering productivity often depends on travelling considerable distances each day in search of roots and nuts. Since the women have to take their offspring with them during these forays, they incur a significant cost in terms of backload. Lee (1982) suggested that !Kung women managed their birth spacing so as to minimize the load they had to carry: a birth spacing of about four years would mean that a woman never had to carry more than one child at a time.

This idea was explored in more detail by Blurton-Jones and Sibley (1978) who modelled the loads that women foraging for a staple resource would have to carry with different interbirth intervals. They pointed out that the woman is not merely having to carry the child (or children) but also the food for herself and her family that she has just gathered. They were able to show that not only was the mean load minimized with a four-year interval but so was the variance. In other words, the mother's back load remained more even over time with a four-year interval than with any other birth spacing.

Blurton-Jones (1987) went on to test further predictions from this analysis. He was able to show that interbirth intervals declined just as would be predicted when the women did not have to forage for all their food (e.g. when they were resident at a cattle post and could obtain a significant proportion of their food from the local store).

While the !Kung data seem to fit the predictions of the backload model remarkably well, it is clear that these analyses cannot be generalized to other human populations in their existing form. When Blurton-Jones *et al.* (1989) applied the models to another hunter–gatherer population (the Hadza of western Tanzania), they found a number of significant differences. Hadza mothers, for example, wean their children up to a year earlier than !Kung mothers. Since suckling is an important mechanism delaying the return of menstrual cycles in all mammals (including humans), it seems likely that Hadza women would have shorter interbirth intervals (though this has yet to be confirmed). Detailed analysis revealed that Hadza and !Kung biology differ in a number of important respects that have a direct bearing on the model as originally formulated. One of these is the extent to which the Hadza women are able to leave their young offspring at camp with older siblings or other relatives as minders in a way that is not possible for !Kung women. In addition, ambient temperatures are much lower in the savannah woodland habitats of the Hadza than in the desert habitats of the !Kung: Hadza women were thus subject to much less heat stress than !Kung women and could travel further (and with more freedom) in consequence. Finally, as a direct consequence of the generally more benign environment, Hadza children are able to forage for themselves in the vicinity of camp, and so are able to obtain a significant proportion of their nutritional requirements at a much earlier age than !Kung children. All these factors significantly reduce the costs of rearing children for Hadza women, thereby making it possible for them to have more children if they so wish.

My last example of ecological adaptation concerns the Tibetans. The Tibetans are among the few human cultures in which polyandry (one woman married simultaneously to two or more men) is considered a socially acceptable form of marriage. Most human societies are either monogamous or polygamous (several women married to one man) in their marriage arrangements. A biologist's natural reaction to such an observation is to ask whether the Tibetans represent a kind of evolutionary error (a non-adaptive strategy that has appeared, as it were, as a natural mutation but which will eventually be driven to extinction under the pressure of natural selection) or a genuinely adaptative strategy. Is there some feature of Tibetan ecology that is in some way different to that occupied by most other human groups which predisposes the Tibetans to polyandry?

Detailed ethnographic studies have revealed that polyandrous marriages are typical only of one group within Tibetan society, namely those families that own their own land which they can pass on to their children. Landless serfs who farm land rented from larger landowners almost without exception adopt monogamy as their preferred marriage pattern. A number of studies (e.g. Goldstein 1971; Crook and Crook 1988) have argued that polyandry has been adopted as a mechanism to prevent the dispersal of large estates under conditions where good agricultural land is in very short supply and most estates are already close to the minimum required to support a single family. By ensuring that all the brothers marry a single wife, the estate is preserved intact; the total number of children born within the family does not exceed the estate's ability to support them, while at the same time sufficient labour to manage the estate is assured. Moreover, the brothers can be counted on to co-operate since they have a common interest in the offspring born to the wife (both through direct paternity, but also through kin selection!). In fact, this, in essence, is basically the explanation that the Tibetans themselves give; indeed, in doing so, they commonly point to individual cases of spectacular disaster when families have failed to ensure that there is only one marriage in each generation.

Durham (1990) has taken the analyses one stage further by simulating the consequences of different marital strategies under the demographic conditions prevailing among the Tibetan peoples living along the Himalayan margin in northern India. These simulations show that although individuals might have more descendants in the short term if they encouraged more of their children to marry, all alternatives except a single (polyandrous) marriage per generation ultimately result in the family's extinction after as few as nine generations.

Crook and Crook (1988) have pointed out that the fundamental driving force behind this situation is the relative poverty of the marginal high altitude habitat occupied by the Tibetan peoples. Although there are undoubtedly many different components to the cultural system of the Tibetans, it nonetheless remains clear that the kind of subsistence ecology which they are forced to adopt is the driving force behind their adoption of polyandry and the principle of a single marriage per generation. This explanation is given further support by the fact that as soon as conditions appear that make it possible for the brothers in a polyandrous marriage to pursue their own livelihoods, then they marry monogamously. Thus, monogamy has become increasingly common in areas like Ladakh following the introduction of alternative sources of income (petty government employment, expedition guides/portering, employment elsewhere, etc).

Mating strategies

While the role of environmental factors in evolution is widely recognized, it is often not appreciated that factors intrinsic to the animals themselves can play an important role in driving evolution. Not everything an animal does is related to some problem of survival in an inhospitable environment: competition among members of the same species for access to mates (usually termed *sexual selection*) has long been recognized as an equally important force (indeed, some biologists argue that it has been *more* important than environmental selection in the evolution of species differences).

Evolutionary theory assumes that, all other things being equal, individuals will behave in such a way as to maximize the numbers of offspring they are able to sire. The theory of sexual selection expands this by suggesting that competition for access to mates will result in some individuals gaining more matings than others do. In some cases, this will be because some individuals can compete more effectively than others (for example, by driving off competitors or by keeping rivals away from their mates): this usually gives rise to *intrasexual selection* (selection for certain characters as a result of competition between members of the same sex). In other cases, it will be because some individuals are more attractive to members of the opposite sex (usually because they are able to offer something that will enhance their mates' abilities to reproduce): this involves *intersexual selection*. What is usually known as *Bateman's Principle* suggests that the sexes will differ in their reproductive strategies as a consequence of the different contributions they make to reproduction. This will be especially conspicuous among mammals due to internal gestation and lactation: these strategic evolutionary effects mean that, in general, females will be more cautious in committing themselves to matings than males and that females will place a higher premium on the rearing component of reproduction compared to males (who will tend to emphasize mating whenever they can). This will result in males competing among themselves for matings, while females should show a marked preference for males who can invest in their offspring. (For a recent review of some of the issues, see Rees and Harvey 1991.)

Evidence that males have a greater variance in reproductive success than females is generally considered to be adequate evidence for sexual selection in any species. The greater the disparity in the variances in lifetime reproductive success for the two sexes (irrespective of which sex has the greater variance), the greater the opportunity for sexual selection (and the more likely it is that this variation is itself the product of sexual selection). At a finer level, we might expect to find that, when reproductive success varies widely across males, there will be a correlation between male reproductive success and some aspect of male status: depending on whether the evolutionary process arises from intersexual or intrasexual selection, these could be features as different as physical prowess, economic wealth, social status, or even willingness to help care for the offspring.

I shall again give three examples of societies in which mating patterns appear to be evolutionarily adaptive in this respect. Two concern traditional societies in the contemporary world, the third a historical European population.

The Kipsigis are an East African pastoralist people related to the Maasai. Although cattle remain an important part of their economy and culture, they have shifted the basis of their economy increasingly towards agriculture since the turn of the century, such that wealthy families now maintain both large herds and substantial land holdings. In her detailed study of these people, Borgerhoff Mulder (1988) found that the variance in the number of surviving children produced by men during their lifetimes was nearly twice as great as that for women: in other words, there was clear evidence for sexual selection. Similar results have been reported from a number of other studies of traditional societies (e.g. Aka pygmies: Hewlett 1988).

In addition, there was a significant correlation between total number of offspring produced and wealth (based on a combined index of cattle and land holdings) in each of six cohorts of men born between 1922 and 1978 (correlation coefficients ranged from $r = 0.22$ to $r = 0.90$, all significant at the $P = 0.05$ level). These results held up even when the data were reanalysed in terms of the number of children per year of married life.

The main reason why Kipsigis men differed so much in reproductive output was that the number of wives they had varied considerably. Some men had as many as six wives, others only one. Borgerhoff Mulder (1988) found that there was a significant correlation between the number of wives a man had and his wealth. The more cattle he had, the more wives he could afford to buy (bridewealth must be paid to the bride's family), and the more land he owned the more wives he could afford (each wife has to be provided with her own garden plot on which to grow the food for her children). It seems that not only were wealthy men able to purchase more wives, they were also more attractive as husbands because they could provide their wives with better conditions in which to rear their children. Thus, both intersexual selection and intrasexual selection seem to be at work in this case.

A rather different example in which female choice seems to be even more important concerns the Ache, a group of traditional hunter–gatherers from eastern Paraguay on the borders of Brazil. In collaboration with a number of other colleagues, Hill and Kaplan (1988*a*,*b*) carried out an extensive study of one Ache population in the early 1980s. They found that a male's reproductive success (estimated in terms of both known and putative offspring) was related to his competence as a hunter. The better hunters (as measured in terms of the mean number of kilograms of meat brought back to camp per hour spent hunting) had more offspring than poor hunters matched for age.

One reason why good hunters had more offspring is that the women preferred them as 'husbands' (marital arrangements are very informal among the Ache). Good hunters are also preferred as extra-marital lovers; indeed, the women attributed three times as many of their 'illegimate' offspring (i.e. those they believed not to have been sired by the male they were living with at the time of conception) to good hunters as they did to the poor hunters (Hill and Kaplan 1988*a*).

Hill and Kaplan (1988*b*) argued that men's hunting activities make an important contribution to the women's ability to reproduce successfully (the products of hunting are divided equally among all the households in the group), so they are particularly concerned to keep good hunters in their group. In addition, the women incur a potentially serious cost if these individuals leave: because other men are unwilling to contribute the products of their hunting to children whose fathers are no longer members of their group, infanticide is common. A child's chances of surviving to the age of 15 years are significantly lower if its father has died or left the group (57 per cent) than if the father is still living in the group (81 per cent) (Hill and Kaplan 1988*b*). By distributing their favours disproportionately in favour of good hunters, the women ensure that these males are more likely to stay with the group than to wander off to join another group elsewhere. Not only do these males stand to gain more offspring through both licit and illicit matings if they stay in the group, but they will also ensure that more of their offspring survive to adulthood.

The Ache offer a strong prima-facie case for female choice at work: essentially, the women are doing their utmost to force the males to play the game the way they want it played—not without some success. However, this study also provides us with a nice example of how difficult it can be to test evolutionary hypotheses in these cases. Although there is good evidence for sexual selection at work here (males have higher variance in lifetime reproductive success than females, and there is a correlation between male quality (as measured by hunting success and reproductive performance), the intriguing question from an evolutionary point of view concerns the evidence for females manipulating males. In order to substantiate this claim, however, we need to show that females who excel at persuading good hunters to stay actually do have higher fitness (as measured by the numbers of surviving offspring) than females who are less persuasive. This would probably be quite difficult to test directly: we would probably have to test it at the group level by asking whether groups of women who were on average more persuasive had higher fitness than groups of women who were less persuasive.

What makes this example particularly interesting from the evolutionary point of view, however, is the fact that the good hunters themselves need not in fact actually have to do better in reproductive terms than poor hunters: they merely have to *think* that they are doing better. From an evolutionary point, it makes no difference whether or not the women's claims about which males are the fathers of their offspring are true, providing the men take their word for it. In other words, there is scope here for the females to manipulate the good hunters into thinking they have sired more offspring than they actually have.

Although we would expect the men to take whatever steps they can to ensure that the offspring attributed to them really were sired by them, there is an important asymmetry operating in the women's favour: they are much more likely to know who fathered each of their children than the men are. At the very least, a woman will know whom she mated with at around the time of conception, whereas as a man will never know for sure whether or not he was the only one to have mated with the woman at that time. This asymmetry itself has important general implications for paternal investment (males' willingness to invest in the business of rearing offspring at the expense of further sirings), and we shall return to it in the next section.

My last example of human mating strategies is drawn from the historical population of the Krummhorn region in Friesland, northwest Germany, during the eighteenth and nineteenth centuries. The parish records of this population have provided us with a rare insight into the long-term consequences of people's reproductive strategies. Voland (1990) examined the numbers of children produced by men of different socio-economic status who married during the period 1720–1750 and found that farmers produced about as third as many children again as small-holders and day labourers. In addition, while smallholders initially had lower completed family sizes than the landless labourers, their long-term success over the following century was greater because more of their children married and fewer of them emigrated. Thus, wealth and land holding had a significant impact on the men's abilities to leave descendants. It is important to bear in mind that none of these groups are particularly wealthy: there are no landed nobility in the sample, only landed and landless peasants.

These figures are all the more impressive when you examine the long-term consequences of these reproductive differentials. Voland (1990) determined the number of surviving descendants that each man actually had over the course of the century following his marriage. The differences were striking and cumulative. Although farmers had only about 10 per cent more offspring than the average for the population 10 years after marriage, their relative success rose so steeply that, 90 years later, they had 190 per cent more living descendants than the average.

These figures allow us to see the longer term consequences of short-term success: by being able to reproduce more successfully, the original 19 landed farmers (representing just 10.3 per cent of the men who married during the sample period) were able to double their representation in the population over a period as short as three generations. If their rate of growth continued at the same pace, their descendants would have come to represent a full 30 per cent of the population after another 100 years. In effect, the genes of the poorer economic groups were being inexorably squeezed out of the population. Notice that this does not mean that everyone was gradually becoming a farmer: the available farm land was limited and farms could not be progressively divided among successive generations of offspring without becoming too small to be economically viable. Rather, it seems that the surplus descendants of the original 19 farmers were gradually percolating down through the social strata to take over the roles of smallholder, tradesmen, and landless labourer.

Voland's analyses of these eighteenth- and nineteenth-century parish records also suggest that female choice may be an important factor at work here. Voland and Engel (1990) were able to demonstrate that women married preferentially up the socio-economic scale, in addition, those women marrying up the social scale did so earlier than those marrying either down the social scale or into the social class they were born into. Women marrying into the wealthiest group (the farmers) were able to maintain a more or less constant completed family size (measured as the number of children surviving to adulthood) irrespective of the age at which they married, whereas those marrying into the lower socio-economic strata generally experienced a marked decline in completed family size with age of marriage.

Voland and Engel (1990) interpret these data as suggesting that women of all socio-economic classes held out for opportunities to marry into the wealthiest economic groups. With male preferences being for the young women because these have the highest reproductive value (future expectation of producing children), the men from the wealthier classes are able to select younger brides. Once past the preferred age, however, the women's choices are severely constricted: if they have not found a wealthy husband by their mid-twenties, they are forced to settle for less desirable marriages with men of lower socio-economic status.

In this case, the advantages of marrying into the wealthier classes in society seems to be that the men have more resources to invest in the woman's offspring. Farmers have sufficient spare income to pay for doctor's bills and a good education, as well as providing inheritances for their sons and daughters when they in their turn marry. The evidence from Krummhorn suggests that the women value these attributes and, within the constraints imposed by the social and cultural milieu in which they live, endeavour to select their future husbands on this basis. This last finding highlights the importance of parental investment in human reproduction, and it is to a consideration of this that we should now turn.

Parental investment

We can view an organism as having a limited quantity of time and energy available to partition between the various components of reproduction. In this respect, the major contrast lies between the amount of effort devoted to mating and that given to parental investment (essentially any resource devoted to an offspring that improves its chances of surviving to adulthood and reproducing in its turn). By definition, the effort devoted to parental care of offspring necessarily reduces the stock of reproductive effort available to invest in another mating. In species that have prolonged periods of parental care (as all mammals do), parental investment is a particularly important component of the organism's reproductive strategy.

Nowhere is this more true than in humans where the period of parental investment extends from the moment of conception right through until the onset of reproduction (perhaps as late as 20 years of age in some populations). Indeed, in most human populations, parental investment in their offspring probably continues at low levels throughout most of the parent's lifespan, ending with a final flourish in the form of inherited wealth willed to their children (at least in those economies in which it is possible to accumulate wealth).

There are a number of major aspects of parental investment theory that I shall not consider here. One, for example, is the impact that paternity certainty (a father's certainty that his wife's offspring really are his) will have on a male's willingness to invest in a woman's offspring (see for example the Ache case discussed in the preceeding section). Instead, I shall focus on just two aspects of parental investment that seem to be especially prominent in humans. One is the transfer of wealth from one generation to the next in the form of inheritances; the other is infanticide. Paradoxically perhaps, infanticide is, from a biological point of view, simply a decision about whether or not to invest in a given offspring: it is no different in qualitative terms to the decision about when a mother should wean her infant on to solid food.

The Kipsigis whom we discussed earlier have a subsistence economy that is partly based on cattle and partly on smallholdings. Cattle, as we noted above, are used to buy wives and play a relatively small part in the actual subsistence economy of the people themselves most of the time. In contrast, a family's land holding is very important since it provides each wife in the family with the necessary resources to rear her children (as well as the surpluses she needs to purchase clothing, medicine, and other modern necessities for them). Recall that Borgerhoff Mulder (1988) found that a man's lifetime reproductive success correlated with his family cattle holdings, whereas his wives' reproductive success correlated with his land holdings, but not vice versa. It should be no surprise to find that the Kipsigis allocate their wealth in just such a way as to maximize the benefits from each type of resource: while the cattle are given to the sons for their exclusive use, the family's land holdings are handed on to the sons for the exclusive use of their wives (Borgerhoff Mulder 1989). In other words, each resource is given to the sex which can make most profitable use of it in terms of turning the resource into grandchildren.

Notice, however, that land holdings are not given to the family's daughters, even though we might suppose that this would be a surer way of guaranteeing that the family's wealth is invested in biological grandchildren. (Remember that parents can always be sure that a daughter's children carry their own genes, whereas they can never be 100 per cent sure that a daughter-in-law's children were sired by their son.)

Two possible explanations might account for the parent's insistence of channelling all their wealth through their sons. One is that it is simply a by-product of a male-biased culture, an epiphenomenal consequence of the nature of the social system, just as Durkheim might have argued. However, if that is so, then it might just as easily constitute a possible case of memic evolution within a memic universe: investing everything via sons may be a rule that meshes better with the other male-biased rules in that kind of society. (Notice that this is not to deny Durkheim's explanation here, but rather to provide an evolutionary explanation for it.) The other is that other features of the socio-ecological system constrain parents into behaving in this way. This might be a consequence of the fact that patrilocality (i.e. female-transfer exogamy in which wives move to live in their husband's villages) results in married daughters living too far from the parental home to make economic use of the parental fields: it would, after all, be extremely inefficient for the wife to have to travel 20–30 kilometres to work in her fields. Some circumstantial evidence to support this suggestion comes from the fact that daughters living near their parents do regularly work their parents' fields.

Clearly, there is no way in which we can decide which of these options is the correct explanation in this case. It should, nonetheless, remind us that our job is not yet done when we have identified a possible function for a behaviour: we need to return to the field and collect the necessary data to discriminate between the alternative hypotheses.

Let me turn now to the curious phenomenon of infanticide. Clearly, we find the thought of parents killing their offspring particularly repugnant. Yet it is far from rare, having been documented in almost every culture that has ever been studied or for which historical records exist. This is not to suggest that human parents routinely kill off their offspring: if that were the case, the human species would have become extinct thousands of generations ago. However, it is clear that parents do sometimes withdraw parental care soon after birth in some instances. The two cases I shall discuss here concern the Eskimo and the same peasant population from northwest Germany that was discussed earlier.

Voland (1984) analysed the survival rates for newborn babies in the parish of Leezen over the 150-year period 1720–1869. He found a striking difference in the survival rates of sons and daughters across the social classes. Among the peasant farmers, sons suffered about 16 per cent mortality during the first year of life, compared to 23 per cent for daughters; in contrast, among the landless day labourers and tradesmen, the reverse occurred (sons suffered about 22 per cent mortality, while daughters experienced only 15 per cent). Voland argued that these contrasts in apparent preference for one sex of offspring reflected the long-term marital and reproductive opportunities available to children in the different social classes. For landholders, it was important both to have a surviving son to inherit the family estate and to have enough labour to work the fields. Because farmer's sons were desirable catches for everyone's daughters, they also had the pick of the girls in all social classes. One inevitable consequence of such behaviour is a surplus of unmarried (and unmarriagable) daughters in the highest social class. Since a daughter's reproductive success depended on finding a son-in-law with a farm to inherit, the farmer's seem predisposed to undervalue daughters. In contrast, because daughters could marry up the social scale (and thus greatly improve their reproductive prospects, as we saw in the preceeding section), the lowest social classes were equally predisposed to favour daughters at the expense of sons (whose opportunities for marriage and reproduction were very much poorer by comparison).

Notice that the differences in mortality rate are not enormous: wealthier parents did not get rid of all their daughters any more than labourers disposed of all their sons. In fact, the differences seem to reflect the marriage opportunities of the two sexes and the local economy peculiar to this population. In contrast, the peasant farmers of the Krummhorn region further to the west pursued the reverse strategy: their daughters had surprisingly low mortality rates in comparison to their sons. Voland and Siegelkow (1990) argue that this reflects an important contrast in the economies of the two regions of Germany. In Leezen, landholdings were relatively large and could be divided between several sons without endangering their economic viability. In contrast, land was at a premium in Krummhorn, with most farms at the limits of their viability; as a result, it was important that each farm be inherited intact by a single son in each generation. Large families containing several sons were likely to be a recipe for bad feeling and resentment.

This is clearly reminiscent of the situation faced by the Tibetan peasantry that we considered earlier, and we might ask why the citizens of Krummhorn preferred to allow their sons to die rather than opting for polyandry as a way of ensuring the viability of their estates. At least part of the explanation has to lie in the cultural milieu: all forms of marriage other than monogamy were strictly forbidden in Europe by the Christian tradition, whereas there were no such strictures among the many faiths prevalent in the Indian subcontinent during the formative period in the history of Tibetan society and culture.

Here, then, is an interesting contrast in behaviour that begs explanation. Of more interest, perhaps, is the fact that at least one likely explanation for the difference gives a significant role to cultural factors that are independent of the effects of biological variables. Once again, it seems that the dual inheritance theory of memic evolution intrudes into the biological situation.

My last example concerns infanticide of very much more serious proportions. Irwin (1989) collated the demographic data from various censuses of Eskimo populations carried out during the early part of this century and found that the sex ratios of offspring varied significantly away from the 50:50 we would normally expect: in some populations, there were as few as 50 girls for every 100 boys in the infant age classes. Yet in almost all these populations, the adult sex ratio favoured women (typically averaging about 120 women for every 100 men).

When he plotted the infant sex ratio against the mean temperature for the location at which the population lived, Irwin found that infant sex ratios became increasingly male biased as the mean temperature declined. Irwin suggested that the reason for this lay in the fact that males suffered exceptionally high mortality rates while out hunting. In the extreme environments under which the Eskimos live, a woman needs a man to hunt for her or she and her children will starve during the winter. She cannot herself hunt during winter since doing so would place both her own life and those of her children at too much risk. In the most northerly populations living under the harshest climatic conditions, the mortality rate of men out hunting is so great that, on average, a woman needs two husbands to see her through her reproductive life. Irwin (1989) argued that, in order to ensure that each reproductively active female has two men available as husbands, the sex ratio has to be biased to the requisite extent in infancy. Surplus women lacking a provider of their own are simply too great an economic burden on the rest of the community.

Although Irwin's argument here seems very plausible, it remains no more than a hypothesis that fits the facts. In order to clinch the explanation, we would need to be able to show that male mortality rates (or male life expectancy) varied in inverse proportion to the infant sex ratios (i.e. that the highest male mortality rates occurred in just those habitats where the sex ratio was most skewed in favour of males). Irwin does not have those data. However, one study of Swedish demographic data from the nineteenth century does provide a prima facie case for just such a relationship: Johansson (1984) gives data showing that both female infant mortality (relative to that for males) and adult male mortality (to marriagable age) increased as populations became more industrialized. An anticipated surplus of adult females seems to be offset by reducing the numbers of female infants that survive the first few months of life.

Once again, we might ask why alternative strategies were not adopted to make better evolutionary use of the surplus population. One way of absorbing the surplus of women would, of course, be for men to become polygamous. The cultural (memic) restraint imposed by the Christian Church seems at least one likely explanation for the persistence of monogamy in the Swedish case. Equally however, the limited wages earned by factory workers may have been insufficient to support more than a single wife and her children. In contrast, it seems very likely that, for Eskimos inhabiting marginal habitats on the Arctic Circle, polygamy is out of the question: a single male simply could not provide enough food for two women and their children during the critical winter months when the family is totally dependent on the man's hunting activities. Again, in this case we would like to see evidence from hunting rates (and male mortality rates on the ice in relation to hunting effort) to confirm that a male really is incapable of provisioning more than one wife.

A case of cultural evolution?

I have, on several occasions, suggested that some aspects of human behaviour (or at least the cultural rules that govern human behaviour) might reflect instances of memetic adaptation (i.e. that the rule is adapted not to some aspect of the biological reproduction system but rather to the meme's own survival within the network of rules in which it finds itself). Let me conclude with one final example that seems to me to constitute a genuine case of memic evolution. It concerns the evolutionary history of the teddy bear and it is an evolutionary story to which almost all of us have contributed in our own small way.

Following its 'invention' in the early 1900*s*, the teddy bear has undergone significant change in its shape. Hinde and Bardon (1985) measured the features of teddy bears on display in the Folk Museum in Cambridge (UK), as well as those of a sample of bears on display in a Cambridge shop. They found that the bears' heads became progressively more rounded with time: the depth of the muzzle became shorter (relative to the height of the head) and the position of the eyes in the vertical plane became lower with time. In other words, the shape of the bear's head shifted progressively through time from being bearlike in the original models to a form more reminiscent of the head of a newborn mammal infant.

Hinde and Bardon (1985) argue that this is a consequence of the manufacturers' progressive response to people's preference for head shapes that resemble the domed foreheads and large eyes of newborn babies, itself an instinctive feature of human behaviour. Whether or not this preference for infantile shapes really is instinctive is not important: it merely provides a basis for the selection pressure that has been acting on the manufacturers (whose behaviour is almost certainly not under any kind of genetic control). In this context, the buyers' 'innate' preferences have the same logical status as the mean temperature of the habitat or the distribution of food patches.

The critical point is that the manufacturers' response is being driven by the sales success of different models: it seems likely that as the design team sits down to consider next year's production models, they opt for those lines that have sold best the year before. They themselves need have no preconceptions about how bears should look: they merely try designs that they think people will buy, some of which sell and some of which get left on the shelves. In effect, the concept of an 'ideal' bear in the manufacturer's mind (surely, an indisputable example of a meme) is changing through time as a direct result of the selection forces being exerted on it (in this case, the purchasing habits of people who buy teddy bears). Each new design is a 'mutant' meme that survives for a while and then goes extinct.

Once again, of course, to be really convinced we would like to see the sales figures. If we could show that, at each stage in the historical process, there was a preference for the bears with the roundest heads and that this preference declined progressively with time as the bear's heads approached the 'ideal' shape, then we could assert with confidence that we have a case of a meme evolving under selection in a purely memetic universe.

References

Barash, D. (1982). *Sociobiology and behaviour.* Elsevier, New York.

Blurton-Jones, N.G. (1987). Bushman birth spacing: direct tests of some simple predictions. *Ethology and Sociobiology*, 8, 183–203.

Blurton-Jones, N.G. and Sibley, R. (1978). Testing adaptiveness of culturally determined behaviour: do bushman women maximise their reproductive success by spacing births widely and foraging seldom? In *Human behaviour and adapation* (ed. N. Blurton Jones and V. Reynolds), pp. 135–58. Taylor & Francis, London.

Blurton-Jones, N.G., Hawkes, K., and O'Connell, J.F. (1989). Modelling and measuring costs of children in two foraging societies. In *Comparative socioecology* (ed. V. Standen and R. Foley), pp. 367–90. Blackwell Scientific, Oxford.

Borgerhoff Mulder, M. (1988). Reproductive success in three Kipsigis cohorts. In *Reproductive success* (ed. T.H. Clutton-Brock), pp. 419–35. University of Chicago Press.

Borgerhoff Mulder, M. (1989). Reproductive consequences of sex-biased inheritance for the Kipsigis. In *Comparative socioecology* (ed. V. Standen and R. Foley), pp. 405–27. Blackwells Scientific, Oxford.

Boyd, R. and Richerson, P. (1985). *Culture and the evolutionary process.* University of Chicago Press.

Cavalli-Sforza, L.L. and Feldman, M.W. (1981). *Cultural transmission and evolution: A quantitative approach.* Princeton University Press.

Crook, J.H. and Crook, S.J. (1988). Tibetan polyandry: problems of adaptation and fitness. In *Human reproductive behaviour* (ed. L. Betzig, M. Borgerhof Mulder, and P. Turke), pp. 97–114. Cambridge University Press.

Daly, M. and Wilson, M. (1983). *Sex, evolution and behaviour* (2nd edn). Wadsworth, Belmont, CA.

Dawkins, R. (1976). *The selfish gene.* Oxford University Press.

Dunbar, R.I.M. (1982). Adaptation, fitness and the evolutionary tautology. In *Current problems in sociobiology* (ed. King's College Sociobiology Group), pp. 9–28. Cambridge University Press.

Durham, W.H. (1990). *Coevolution: Genes, culture and human diversity.* Stanford University Press.

Durkheim, E. (1915). *Elementary forms of religious life.* Humanities Press, New York.

Finch, V. and Western, D. (1977). Coat colours in pastoral herds: natural selection or social preference? *Ecology*, **58**, 1384–92.

Goldstein, M.C. (1971). Stratification, polyandry and family structure in central Tibet. *Southwestern J. Anth.*, **27**, 64–74.

Hamilton, W.D. (1964). The genetical evolution of social behaviour. I, II. *J. Theor. Biol.*, **7**, 1–52.

Hewlett, B.S. (1988). Sexual selection and paternal investment among Aka pygmies. In *Human reproductive behaviour* (ed. L. Betzig, M. Borgerhoff Mulder, and P. Turke), pp. 263–76. Cambridge University Press.

Hill, K. and Kaplan, H. (1988*a*). Tradeoffs in male and female reproductive strategies among the Ache: part 1. In *Human reproductive behaviour* (ed. L. Betzig, M. Borgerhoff Mulder, and P. Turke), pp. 277–90. Cambridge University Press.

Hill, K. and Kaplan, H. (1988*b*). Tradeoffs in male and female reproductive strategies among the Ache: part 2. In *Human reproductive behaviour* (ed. L. Betzig, M. Borgerhoff Mulder, and P. Turke), pp. 291–306. Cambridge University Press.

Hinde, R.A. and Bardon, L.A. (1985). The evolution of the teddy bear. *Anim. Behaviour*, **33**, 1371–3.

Irwin, C. (1989). The sociocultural biology of Netsilingmiut female infanticide. In *The sociobiology of sexual and reproductive strategies*, (ed. A. Rosa, C. Vogel, and E. Voland), pp. 234–64. Chapman & Hall, London.

Johansson, S.R. (1984). Deferred infanticide: excess female mortality during childhood. In *Infanticide: Comparative and evolutionary perspectives* (ed. G. Hausfater and S.B. Hrdy), pp. 462–85. Aldine, New York.

Lee, R. (1982). *The !Kung San: Men, women and work in a foraging society*. Cambridge University Press.

Lumsden, C.J. and Wilson, E.O. (1980). *Genes, mind and culture*. Harvard University Press, Cambridge, MA.

Maynard Smith, J. (1989). *Evolutionary genetics*. Oxford University Press.

Prout, T. (1969). The estimation of fitness from population data. *Genetics*, **63**, 949–67.

Rees, J.A. and Harvey, P.H. (1991). The evolution of animal mating systems. In *Mating and marriage* (ed. V. Reynolds and J. Kellett), pp. 1–45. Oxford University Press.

Voland, E. (1984). Human sex ratio manipulation: historical data from a German parish. *J. Hum. Evol.*, **13**, 99–107.

Voland, E. (1990). Differential reproductive success within the Krummhorn population (Germany, 18th and 19th centuries). *Behavioural Ecol. Sociobiol.*, **26**, 65–72.

Voland, E. and Engel, C. (1990). Female choice in humans: a conditional mate selection strategy of the Krummhorn women (Germany, 1720–1874). *Ethology*, **84**, 144–54.

Voland, E. and Siegelkow, E. (1990). Comment on 'Sieff: Explaining biased sex ratios in human populations'. *Curr. Anth.*, **31**, 42–3.

4
CULTURAL ADAPTATION
Howard Morphy

Introduction

For much of the twentieth century the concepts of culture and adaptation have been almost equally neglected by British social anthropologists. In both cases the origins of this neglect can be traced to the rejection of the nineteenth century evolutionist paradigm and its replacement by the paradigm of structural functionalist anthropology, which in Britain was associated with the switch from a historically based subject to one that became primarily focused on the study of existing societies through fieldwork. Culture was rejected because it was too closely associated with the rag bag check list of traits that were used in a somewhat mechanistic way to order cultures in hierarchical sequences, reflecting nineteenth century concepts of progress. Adaptation which through the concept of natural selection was integral to Darwinian theory was rejected on similar grounds since it too, when applied to human beings, was associated with an ideology of progress. Moreover the switch of focus to social relations and the neglect for the most part of the material and economic dimensions of human existence made the concept appear a less relevant one. With hindsight there was certainly a place for a coherent theory of adaptation within the needs-based functionalism of Malinowski and the organic analogy of Radcliffe-Brown.[1] In Malinowski's case function may have been assumed to have a connection with the satisfaction of biological needs, but that proposition was left largely unexplored. In Radcliffe-Brown's case the organic analogy operated within an artificially constructed domain of social relations cut off from the material world (but see Burnham 1973, p. 98 on the subsequent relationship between systems theory models of adaptation and functionalist teleology). The fact that neither developed a theory that included adaptation is in part a sign of the arbitrary divisions within anthropology that were imposed by the structural functional paradigm.

American anthropology did not undergo the same extreme paradigm shift as British anthropology, for although Boasian anthropology was equally antipathetic towards nineteenth century evolutionism, the debates over evolution and culture change remained as part of the agenda of American anthropology. In American anthropology the role of culture as an adaptive system has been a continuing theme, as has been interest in the related areas of cultural ecology and evolutionary theory in the writings of White, Kroeber, Steward, and subsequently Sahlins, Service, Harris, and Rappaport. In Britain interest in the topic was carried forward almost single-handedly by the work of Daryll Forde, and occasional landmark studies, such as Evans-Pritchard's (1940). *The Nuer* and to a lesser extent the work of Richards (1939) on Bemba agriculture and economy, shone out like beacons in a sea of neglect (see Ellen 1982, p. 29). It was not until the 1960s and 1970s that a renewed interest in ecology developed in British anthropology, in particular in the study of hunting and gathering (Woodburn 1968), pastoral (Dyson-Hudson 1966; Ingold 1980), and subsistence agricultural societies (Ellen 1978; Burnham 1979). The renewal of interest was given added impetus by the French structural-Marxism of the 1970s which had a strong ecological strand built into it (see Godelier 1977; Meillasoux 1981).

Evolution, the environment, and culture

Adaptation I define minimally as the survival of a group in a given environment or its adjustment for survival in a new environment. (I am referring here to the physical environment rather than the social or cultural environment, though there is clearly a sense in which the concept of adaptation might be applied to a group's adjustment to changing political and cultural contexts.) As Sponsel (1986, p. 73) writes:

> The central proposition of cultural ecology is that the ecosystem plays an active role in the daily lives of a human population and interacts to some degree with every component of the cultural system (infrastructure, structure, superstructure). A corollary of this is that culture may be viewed as a system that adapts the population to its ecosystem, i.e. it promotes the survival, maintenance, and reproduction of the population.

To which he adds the important rider: 'this does not mean that culture is only an adaptive system'. Adaptation can be viewed from both a short term and a long term perspective. In the short term it can refer to the ways in which a population is able to modify its behaviour in order to cope with changing climatic factors, seasonal variation, and environmental diversity. It is the potential for variation within certain parameters in a given human culture. In another and more determining sense it refers to a population's ability to maintain or increase its fitness in a given environment, i.e. an adaptive strategy is something that enables a population or perhaps species to survive on a long-term basis. In this context I am applying the concept of fitness at the population or species level rather than at the level of the individual. The difference is important. A population of well-adapted individuals may not be a well-adapted population since rapid short-term growth may overload the resource base in the longer term.[2]

In general I will be adopting an evolutionary perspective that has its origins at least as early as the works of A.R. Wallace (1870). From this perspective culture is integral to the way in which human beings articulate with the environment and accommodate to changes in environmental circumstances. Culture, consequently, is part of the process of the reproduction of human beings as components of viable populations. It is of course one thing to be conscious of the fact that culture and social organization reflect the needs of human beings in particular environments by enabling the provision of food and shelter appropriate to the time of year and the size of the community, it is quite another to argue that the particular form that a society takes is determined by the satisfaction of those needs.

For heuristic purposes it may be useful to draw a distinction between adaptation in the weak sense and adaptation in the strong sense, to distinguish between the very different role given to the concept of adaptation in different theoretical frameworks. I will argue that the distinction is a relative one rather than a substantive one. Adaptation in the weak sense refers to studies which aim to show a correlation or in some cases a determinate relationship between certain social or cultural variables and features of the environment, or which show how a particular system operates in the context of regional or seasonal change in environmental resources. Classic studies of the first type show the relationship between population and resource base, as in Birdsell's (1953) famous study of the relationship between rainfall and population density in Aboriginal Australia.[3] An example of the second type would be Evans-Pritchard's (1940) study of the Nuer in which he was able to show how the organization of groups articulated with the seasonal ecology of cattle herding. Adaptation in the strong sense refers to studies which aim to demonstrate that adaptation to the environment has been a causal factor in the development of a particular socio-cultural system and in the process of its transformation over time. An example of a strong adaptationist perspective is Harris's (1979) cultural materialism in which he at times comes close to arguing that, given the particular techno-economic system, certain cultural practices, such as the sacred treatment of cattle in India, evolved because they enable the population to survive in a particular environment (Harris 1965). Strong adaptation would also include analyses which apply a criteria of fitness to human cultural practices, such that their presence is assessed on the basis of their contribution to the viability of the population or the genetic fitness of the individual.

The distinction I am drawing here in many respect corresponds with that made by Ellen (1982) between possibilism and determinism as polar opposite ends of a continuum of approaches. Possibilism represents a perspective in which the environment is a passive limiting factor rather than a causal factor (see Ellen 1982, p. 21). Determinism is a perspective in which it is possible to define certain variables, in this case environmental variables, as causal factors. An extreme possibilist position would be that human beings given the appropriate technology and resources can overcome virtually any environmental constraint. The determining effect of the environment is consequently minimal. However such an argument is usually unhelpful precisely because it neglects the particular historical circumstances and the existence of determinant factors such as the technological system, the division of labour, and so on. As Moran (1991, p. 362) writes 'If people can intensively farm the Negev Desert they can also farm the Rio Negro watershed—at an enormous capital cost in both cases'. Likewise in its extreme form environmental determinism is demonstrably wrong because of the enormous range of economies possible in any given environment. In a sense possibilism is a negative argument against determinism which exhibits the theoretical weakness that determinism is designed to counter. Whereas determinism seeks causal relationships, possibilism sees only opportunities, constraints, and limiting factors. The weakness of possibilism is that culture is not a given entity that exists independent of human adaptive processes in an environment. The weakness of environmental determinism is that human action is mediated by and influenced by human cultural systems. The dialectical nature of the relationship between determinism and possibilism and the need for the development of middle ground theories is well argued by Ellen (1982, p. 51):

> To view determinism and possibilism as mutually contradictory simply perpetuates a logical tautology. The environment does indeed set limits on what is logically possible under a given set of conditions. It sets the outer limits for the evolution of any system and the character of their reproduction; and it sets technical limits to developmental processes and developmental cycles, such as the selection of garden sites. But in restricting the options it is also helping to determine the final outcome. Throughout, the effects of the environment are mediated by social relations which bear a particular cultural load; both are the product of an immediately contingent historical sequence. This appears to have been what Forde meant by the 'middle term' (1934, p. 463), and it is this which determines the variety of possible responses to changing conditions. The problem for the anthropologist is often not to discover whether a variable constrains or determines, but to assess the degree and character of its flexibility. And the extent to which a variable constrains or determines is itself dependent on the cultural configuration resulting from those historically prior environmental relations through which it is mediated.

The logic of this argument must be that in the particular cases variables, whether they be environmental features such as soil fertility, cultural factors such as knowledge of crop rotation, or social practices such as birth control, are frequently interrelated in their development. As such they are, on a priori grounds, best conceived of as multiply determined components that exist in the context of the historical trajectory of a regional system. While I accept that many aspects of a cultural system are neutral with respect to adaptation, the fact that the particular form of a cultural practice is in part affected by its integration within an adaptive system is to me the corollary of the influence that cultural factors have on adaptation: multideterminacy works both ways. I also accept the possibility that there will be rare cases in which a cultural practice is strongly determined[4] by its adaptive value. In harsh environments, such as the arctic or the Kalahari Desert, environmental constraints are, relatively speaking, more deterministic in their effect than in more temperate conditions. Given a particular mode of production such as hunting and gathering then one might expect a greater degree of parallel evolution in arctic or desert regions (Peterson 1979) than elsewhere. In general, however, I prefer to adopt an intermediary position which sees adaptation as one of the factors which influences the way in which a cultural system develops over time. But it is only one factor operating in conjunction with others and it is unlikely to be the sole determinant of a particular cultural practice. Indeed even in the case of harsh environments where it is possible to compare the development of societies with similar modes of production immense diversity is exhibited.

It is sometimes argued that the difference between an evolutionary perspective, which attempts to see the dynamics of adaptive processes as they are worked out over time, and a more local perspective, in which the analyst seeks to establish correlations and systematic relationships between environmental, ecological, and socio-cultural factors, is a matter of scale. Although differences of scale certainly affect the kind of analysis that is possible, I do not think scale should affect matters of theory. As with most debates involving the relationship between synchronic and diachronic perspectives, the opposition needs to be dissolved: if long-term processes are involved in determining what is presently observable on the ground then the analysis must at least take account of the effect of those long-term processes and be able to model them into the analysis where relevant.

Culture

In this chapter I adopt an ideational definition of culture in which culture is seen as a body of knowledge and ideas that is transmitted by human beings, and objects and actions are seen to be the product of those ideas (see Sahlins 1976). Material culture objects and technological systems are part of that body of knowledge and as such are integrated within the ideational system as a whole. Such a perspective sees individual action as mediated by cultural ideas but does not necessarily deny the individual a creative role in the reproduction of those ideas. Culture is logically prior to action as genotype is to phenotype, but action is integral to the process whereby culture is modified through the process of its transmission. Culture is not a sufficient explanation of human action but is a necessary component of any explanation.

Culture mediates between people and their environment. The adaptive components of a culture include its productive technology and the practices used by people to gain a living and to maintain health and control over population. 'Mode of adaptation' could be simply another way of referring to a technology or a mode of production or some other concept that separates out those components of a cultural system that are necessary for group survival from those that are not. Steward (1955), for example, developed a concept of a cultural core which included those components of a particular cultural system that were most closely related to the exploitation of a particular environment. Apart from the extreme difficulty of any such exercise I would see it as being theoretically flawed. Adaptation must include the wider consideration of the relationship between a population and the environment and the theoretical issues that revolve around that relationship, and it must also include the way in which adaptive practices are integrated within the overall framework of a cultural system. We are concerned with survival not as an abstract concept but as a particular kind of way of being human: it is arguable that the cultural value of a practice is just as important in determining its reproduction as its survival value.

Some people have attempted to develop analytic frameworks which for theoretical and methodological reasons exclude the non-practical aspects of technology and material behaviour from having a determining role, at least in the first instance, in the development of a socio-cultural formation (Harris 1979). It may be a useful analytic exercise to push the material explanation of a cultural practice as far as possible, but as a theoretical position it is fundamentally flawed. This is partly because in a given situation there are likely to be many alternative solutions to a survival problem. But even if a criterion of efficiency is employed in assessing the relative survival value of alternative practices, it will not be decisive in explaining why a particular one is accepted (see Jones 1977 for a particularly stimulating discussion of this issue). A marginal survival advantage may be insufficient to counter the greater cultural value of an alternative. The long-term consequence of adopting a marginally less efficient practice may simply be a slightly lower population density, or a working day that is longer by two minutes! In the majority of cases the adoption of a particular practice and the consequences for the historical trajectory of a particular socio-cultural system is likely to be multiply determined. Thus 'technology' as a cultural idea is not easily separated out from the way a particular object, action, or process is integrated within other aspects of the ideational system: ideas about the proper way to do things, the division of labour, 'it is a man's weapon' or a 'woman's weapon', 'it is something that you don't do on Sunday', and so on. Culture may act as a constraint on action as well as providing a potential for action. For this reason Sahlins (1976) has argued that, all things being equal, cultural reason is an underlying determinant of the form of practical action, or as Durham (1991, p. 204) states 'simply put, cultural values can be expected to bias cultural change towards variants that match or "fit" the existing web of local meaning'.

So far I have been talking about an adaptive system as if it is unproblematic, as if it is something that can be easily identified and defined. In many respects an adaptive system, like a cultural system or a social system, is an abstraction rather than a neatly bounded entity. As soon as I mention a system that can adapt itself to different regional environments I pose the problem about the boundaries of the system. How do we draw the line between different variants of an adaptive system and different systems of adaptation. When taking a historical perspective how do we know whether we are dealing with the same system of adaptation adjusting to the changing seasons or with different systems of adaptation for different seasons. We will see shortly that these questions of system definition are not idle ones (see Ellen 1982, p. 78 ff. for a useful discussion).

Although the concept of a cultural system too is problematic, it is useful as a logical type for formulating arguments and for modelling the way in which reality diverges from ideal type. It is particularly so if one believes that culture consists of structured systems of knowledge that influence people's objectives and the direction in which that population evolves.

Culture and society

In discussing the role of culture in human adaptation, it is necessary to consider, albeit briefly, the epistemological problem of the relationship between the key concepts of culture, society, environment, and the gene. The debates over adaptation have often polarized over the different status given to these concepts in the reproduction of socio-cultural systems, involving respectively, cultural determinism, social determinism, environmental determinism, and in the case of some versions of socio-biology, a form of genetic determinism. Or culture can be framed as analogous to genetic material (in the case of Dawkins's (1982) memes), as an environmental resource, or as an epiphenomenon of society.

Theoretical debates in American anthropology have sometimes been structured around the issue of whether culture or material adaptation to the environment is the basic determinant of the form of socio-cultural systems. 'The alternatives in this venerable conflict between utilitarianism and a cultural account may be broadly phrased as follows: whether the cultural order is to be conceived as the codification of man's actual purposeful and pragmatic action; or whether, conversely, human action in the world is to be understood as mediated by cultural design, which gives order at once to practical experience, customary practice and the relationship between the two' (Sahlins 1976, p. 55). British and American anthropology on the other hand have often seemed divided over the utility of the concept of culture itself (see Ingold 1986). The recent books on evolutionary theory by Durham (1991) and Ingold (1986) provide a case in point.

Although the foci of Ingold's and Durham's analyses differ somewhat and there are some substantive differences between their approaches, in many respects the arguments they present are quite compatible. They disagree most profoundly on whether culture or society is the determining level in considering human action in the world. Both see consciousness as a crucial component of the way in which humans can influence biological evolution, but Durham tends to locate consciousness as part of a cultural process whereas Ingold locates it effectively at the level of individuals acting in the context of social relations. Such differences in perspective arise partly because concepts such as culture cover fundamentally different grounds in British and American anthropology. Culture as a concept occupies a potentially crucial position in the debates concerning the relationship between process, structure, and action. Culture can either be seen as a resource that human agents utilize for particular objectives or as something constitutive of the person and integral to the particular form of consciousness of members of a society, and, consequently, as a determinant of their objectives. Thus Durham draws an analogy between genotype and phenotype and culture and behaviour. Culture is reproduced as a system of ideas that is expressed in behaviour—the behaviour itself has meaning only in relation to the idea by which it is understood and which motivates it. Ingold on the other hand sees culture as a resource that enables social action to be carried out. In a significant passage he (1986, p. 262) writes: 'Culture, manifested as behaviour, translates social purpose into practical effectiveness. Whereas interactive behaviour is subservient to the reproduction of culture, the latter is in turn subservient to the social reproduction of life.'

While culture and society can be polarized in this way to correspond with opposite sides of the debate about how human societies reproduce themselves, it is equally possible to define the concepts so that they are both integral to the process. While British anthropologists have at times appeared stubborn in their reluctance to accept the existence of culture as a level of reality in their analyses, American anthropologists have been accused of being idealists in their neglect of the role of social action in the process of the transmission and transformation of ideational systems. With the recent development of more praxis oriented approaches to culture and society there is no longer any benefit to be gained from maintaining the separateness of these positions. Human action in society must be the mechanism whereby ideas affect the world, but this does not make actions in any ultimate sense logically prior to ideas. Both are part of the same process and each is necessary to the existence of the other. If as Ingold implies culture has no role in the definition of human purpose, then social purpose becomes something that arises *sui generis* out of the action of human beings as social beings in the environment. Yet how can cultural knowledge be excluded from the process by which people become aware of themselves as existing in the world as human beings of a particular kind, and how can that awareness not contribute to the kind and form of social purposes that arise? Culture cannot simply be a resource but is something that exists both inside and outside the individual. In his discussion of the nature of consciousness Ingold comes close to recognizing this fact.

Following Giddens, Ingold (1986, p. 207) sees conscious states as the product of the 'reflexive monitoring of action' by a person stimulated by the request to render an account of his behaviour to himself or others. Consciousness develops through the process of social relations and it is the communicative context that enables an individual to get outside himself or herself. 'Consciousness enfolds social relations and unfolds in social relations.' Although Ingold's insightful phrase highlights what must be an important component of the process, it leaves culture, which, at least in the form of languages must be part of the process, as a passive resource outside the equation. Yet just as culture must influence qualitative aspects of social relations so too must it influence the form of consciousness. In order to accommodate culture within the process Ingold has to divide the human agent into two, distinguishing the individual (as the culture bearer) from the person (as the locus of consciousness). 'Thus culture stands, in a sense between the person and the individual; worked on by one, it works on the other'

(Ingold 1986, p. 293). His failure to see culture as an integral part of human consciousness forces this solution.

Environment and genes

The nature of the relationship between culture and genes has been as controversial as that between culture and social relations. Is culture a relatively autonomous system analogous to the genetic code, and if so, is it a characteristic unique to human populations and uniquely a component of human evolution?

Adaptation is concerned with the relationship between a species and its environment, both in the weak sense of articulation with the environment and in the strong sense of evolution to fit the environment. The dominant model for most animal species is that the animals' behaviour expresses the relationship between genetic potential and external environmental conditions. Behaviour involves triggered responses to internal and external stimuli, individual learned knowledge, and knowledge gained from the behaviour of other animals. Frequently the behaviour results in the production of objects which exist outside the body of the animal and which become part of the environment (Dawkins's external phenotype)—objects such as beehives, birds' nests, and beaver dams. Moreover through the process of food acquisition animal species alter the balance of the relationship between species in the outside world. On the whole such products and byproducts of animal action are considered the product of their inherited genetic potential in a given environment rather than of an inherited and transforming body of cultural knowledge. The individual ant or bee is not able to conceptualize the object that it plays a part in producing; rather it plays a role determined by a programme which causes it to act in certain ways in given circumstances as a response to particular stimuli.

Some people have taken the view that human culture is analogous to the external products and learned behaviour of other animals, and that the reification of learned behaviour in the set aside concept of culture creates an artificial difference between humans and other animals. Human culture, it has been argued, could best be treated as the expression of a potential human phenotype in a given environment. However it is difficult to sustain this argument without circularity, for example by including cultural knowledge as an environmental resource. In the same or similar 'natural' environments at different times and in different places human beings have developed fundamentally different responses, and these have resulted in an enormous variety of ways of organizing people and of gaining a living. The individual gains the knowledge that enables him or her to participate and contribute to a particular way of life. That knowledge does not come from within the person, though the capacity to acquire, organize, and act upon knowledge does. A body of knowledge cannot be placed within the individual even though it is individuals who are responsible for its replication and transformation, nor can it easily be placed in the environment, though it does clearly become part of the world in which people exist. An essential component of the concept of environment is that it exists outside the organism, although strict separation is impossible since the organism affects the environment by acting in it. If culture, as the means by which human beings conceptualize the environment, is part of the environment in which humans act, then it is part of an environment which only human beings have access to—in other words it is a specifically human bracketed version of the environment. It would be quite compatible with an ideational definition of culture to argue that animals too have a culture to the extent that they have a body of knowledge that enables them to conceptualize the environment in a particular way. Even if that argument is accepted, human culture is exponentially more complex than that of other animals and is subject to more rapid processes of change and elaboration (see Ingold 1986, p. 359).

The overall position I adopt in this chapter is that human action in the environment is the product of genetic and cultural inheritance. Society and social relations have an existence externally as part of the environment in which individuals grow up and to which they respond, and internally as part of a system of ideas and meaning integral to the way in which the individual gains an understanding of the world and conceives social institutions and relations. There is also a genetic component to social relations, both in the capacity to produce and sustain those relations and in the emotional stimuli which contribute to their formation.

The fact that the different levels of reality identified can in some respects be included within each other, that the environment is something that is both conceptualized by humans and yet includes them within it, is consistent with the fact that we are dealing with a process in which the components of a transforming reality (the existence of human beings in the world) are interrelated. Neither culture nor environment, genes nor social relations are truly independent variables. To adopt a position of multi-determinacy, or co-evolution to use Durham's related concept, is to say that (by definition) the various levels of reality isolated for analytic purposes interpenetrate each other. That does not mean that they cannot also be given relative autonomy. The environment cannot be reduced to human conceptions of the environment even if those conceptions influence the way people act in the environment and in turn contribute to environmental change. The environment (in the sense of the physical world) exists simultaneously inside and outside human conceptualizations of it:[5] it is subject to changes quite independent of human action and will exist beyond human extinction.

Genes, culture, and consciousness

There is a growing consensus among those interested in evolutionary anthropology that human culture has evolved as a relatively independent system of adaptation from human genes. That does not mean that there has been no interaction between cultural and genetic change. In the long term as Ingold has argued 'the total independence of genetic and cultural transmission is difficult to accept, if only because the capacity to acquire cultural information is genetically encoded and not vice-versa'. Clearly the human capacity for culture has a basis in genetic evolution. Moreover there is some evidence that human cultural processes can in turn have an effect on genetic change. Certainly Durham (1991, p. 226) argues that in the case of lactose absorption in adult populations 'The genetic evolution of this rather specialised physiological capacity [illustrates] the important ways in which culture contributes to the evolution of human genetic diversity'. However with a few notable exceptions it has proved difficult to apply criteria of fitness to the interaction between genetic and cultural processes and frequently there seem to be major discontinuities between genetic replication and cultural survival.

Biological adaptation involves natural selection. Those species that survive and those individuals whose genes multiply are the fittest. There is an underlying assumption that evolutionary processes operate to the long-term advantage of a species and enable it to adapt to changing conditions of existence. With the hindsight provided by the fossil record it can be seen that certain species follow evolutionary trajectories that result in their ultimate extinction and replacement by others, while in some cases their phylogenic forebears continue to survive. As Harrison (1991, p. 2) writes 'the critical biological test [of adaptation] is of the displacement, incorporation or extinction of one group by another; and that cannot be tested with one's ancestors'. Advantage in evolution can never be absolute unless it results in the extinction of a competitor and then the victory may be of short duration. In human beings cultural adaptation is certainly one of the ways by which one group of humans gains advantage over another. As with other species survival or increased fitness is often at the expense of others. In evolutionary terms this can make sense, unless the other is necessary to one's own continued existence. In the human case culture and society intervene in this process to influence which adaptations continue to prosper and whose genes proliferate. It is important to emphasize that genetic survival and cultural survival are not necessarily connected. Populations can move from being hunters and gatherers to being pastoralists without any genetic discontinuities, even though subsequently as pastoralists or as hunters and gatherers the populations will be subject to different selection pressures. However, force directed against other cultures can intervene in the evolutionary process. The processes that result in cultural change may also result in the extinction of populations. Colonial expansion has resulted in the destruction of many populations and cultures just as the introduction of Western medicine has arbitrarily increased the members of others.

Processes of cultural change, including colonialism and the use of force, can result in cultural death even if a cultural system is otherwise well adapted to the environment in which the people lived. The original culture may represent a better long-term adaptive strategy in the given environment than the cultural form which replaces it. The replacement of hunting and gathering economies in harsh environments may be a case in point. The environmental, social, and economic costs of maintaining Aborigines in the Western Desert of Australia, or Inuit in the Canadian arctic, as outliers of a modern industrial welfare state is far greater than if they had continued as autonomous hunting and gathering societies. In the case of the Western Desert, the human population was probably at its highest during the hunting and gathering mode of its existence. As Sponsel (1986, p. 82) writes with reference to the Amazon: 'indigenous cultures have millennia of adaptive experience in Amazonia. Perhaps they are developed whereas in Amazonia Western society is underdeveloped.'

Choice and consciousness both play crucial roles in human evolution. Through choice people play an active role in deciding which life to follow and which adaptive strategy to adopt, and through being conscious they do so with some awareness of what the consequences of exercising that choice will be. Choice, however, is constrained not only by the options that are available but also by the political structures and power relations within which the individual making a choice is embedded. Political leaders and elite groups may have a near-monopoly over decision-making such that they rather than the population as a whole or the accumulated sum of its individual members determine the trajectory of the population. Likewise consciousness is limited by knowledge and by the social and cultural factors that structure it.

The importance of human consciousness in decision-making makes it necessary to build both the cultural and the social dimensions of human society into models of human adaptation. Culture provides a resource that is in many respects analogous to the genetic code, except that it is external to the organism itself. Society must be taken into account because human action and human consciousness are factors in both cultural and genetic change. People may consciously influence the process of adaptation by selecting options or by monitoring adaptive systems in order to optimize their chances of survival. And it is possible that human groups or individuals can include long-term factors in this equation. Such strategies may have an effect on genetic evolution when they favour individuals of a particular phenotype over others. In this way human consciousness may indirectly influence biological evolution, since a given choice may differentially affect the fitness of members of the population. Irrespective of the degree of consciousness involved it must be recognized that 'human evolution has taken place within a cultural environment in which many of the selections operating derive directly from the nature of cultural requirements' (Harrison 1972, p. 13). In this paper, however, I am not so much concerned with the way in which cultural and social evolution may effect genetic evolution, but more with the role of social and cultural processes in adaptation.

Adapting to the environment

In this section I look at the relationship between culture and the adaptation of a population to the environment on three different scales: the relationship between a population and the environment within a particular socio-cultural system; synchronic variation in a regional system characterized by environmental diversity; and the development of a regional cultural system over time in a region of diverse environments. I argued earlier that there is no necessary theoretical difference associated with differences in scale and that the results of long-term or regional analysis are relevant to the detailed study of a particular local adaptive system. I acknowledge that the boundaries drawn around any cultural system are inevitably somewhat arbitrary given the complex historical interconnections that all human populations are subject to. Because I am concerned to show the relationship between culture and adaptation to particular environments I choose examples at the local level which demonstrate how a population responds to short-term environmental change. For comparative purposes I look at examples in which cultural continuity can be demonstrated across the regional populations. My methodological justification for proceeding in this way is the need to control one of the variables (culture) as much as possible to see if the impact of another, the environment can be demonstrated. However there is also a theoretical justification in making culture rather than environment the constant, since I would argue that there is no reason a priori why one would expect to find a relationship between environment and culture except in cases where the societies have some cultural components (such as their techno-economic system) in common. There is no need to reargue the case against environmental determinism that was so convincingly put many years ago by Daryll Forde.

Environmental factors can be among the co-determinants of the historical trajectory or the spatial transformation of a cultural system. Indeed the concept of an 'adaptive' system is predicated on this assumption. One of the main reasons for looking at human populations in a long-term perspective must be to attempt to analyse out the relationship between different variables, including the effect that change in environmental or cultural factors has on other parts of the system. I do not imply a monocausal reductionism, a dismissal of history, or a belief in some ultimate level of determinism, but a recognition that changes in certain variables have predictable consequences and that the analysis of human responses to those changes can provide valuable insight into the evolution of socio-cultural systems over time. The variable may be social, for example the strengthening of hierarchy or the intensification of production, or it may be environmental, for example epidemic disease or environmental degradation. Change in the relationship between a population and its resources (as in the states of Highland Burma, see Leach 1954; Friedman 1975) or the movement of a population into a new area (as in the populating of Polynesia, Sahlins 1958), provide opportunities to examine both the sense in which a particular cultural system is an adaptive mechanism and the way in which the environment is a factor in changing the system. First I consider some examples of more local adaptive systems.

The interrelationship between culture, environment, and social organization

It is useful to distinguish between studies which describe a limited correlation between environment and cultural practice, and those in which a more global relationship between environmental factors and the socio-cultural system is posited on theoretical grounds. The former are likely to be positioned more towards the possibilist end of the continuum and the latter towards the deterministic end. There are many studies of the former type—perhaps most common are studies that show the relationship between seasonal factors and group organization (e.g. Thomson 1939; Evans-Pritchard 1940), or soil fertility and subsistence agricultural practice (see, for example, Clarke and Haswell 1971). The more global approaches include ones informed by the theory of ecosystems (e.g. Rappaport 1967) and those based on the assumptions of formal economic theory (e.g. optimal foraging theory, c.f. Smith 1983, 1988). At the most extreme end of the continuum is the cultural materialism of Marvin Harris (1979) in which in the first instance the analyst is encouraged to explain any and every component of the socio-cultural system in terms of its contribution to the material reproduction of the society. It could be argued that the difference between the two types of study lies in part in how far the environmental explanation is pushed, and to what extent other socio-cultural variables are explained by it. The relationship between seasonal factors and patterns of aggregation and dispersion has been well demonstrated in a number of hunter–gatherer and pastoral societies, but it is going a step further to argue that the particular form of group flexibility which is achieved is explained in part by the requirement for flexibility, and a step further still to argue that the form is determined by the need for flexibility. It is unfortunate that the obvious absurdity of the last position (for different societies clearly achieve flexibility in different ways) was not recognized, and that as a consequence people were prevented for some years from exploring the intermediate position.

The seasonal factor in Nuer life

Data on the Nuer have provided the basis for a whole series of studies on the relationship between ecology and social organization. The Nuer are one of a group of peoples in the region of the Upper Nile whose economy is based on pastoralism, supplemented by or combined with the cultivation of crops. An economy with a bias towards pastoralism has characterized the region for at least a millennium. Evans-Pritchard (1940) foregrounded the concept of ecology in his analysis of Nuer society, but he used the concept in a fairly limited way.

The Upper Nile region of the Sudan is a region of swamps and clay soil grasslands intersected by numerous rivers and watercourses. The plains are flat with minor elevations which take on considerable significance at the height of the seasonal floods. In the wet season the combination of rain and the rising of river levels caused by overflow from the Central African lakes into the Nile system results in the inundation of the country. With the end of the wet season the flood waters retreat and gradually the country becomes accessible. Eventually only a few sources of fresh water remain in places where it is possible to graze large herds. The pattern of human life follows this seasonal cycle. In the wet season people move into permanent village sites which are situated on elevated areas of land which are sufficiently large to support the village's cattle in a normal year. As the dry season develops people have more options. For a while the cattle can be herded in cattle camps. These are separate from the village sites where the majority of the population remains to continue with daily tasks, repair the cattle byres, and harvest the final millet crop. The village community spreads out as the dry season progresses into smaller groups of one or a few families in order better to exploit the regional resources. Towards the end of the dry season water (this time its scarcity) again becomes a limiting factor and the people congregate around those most favourable sites of permanent water. Evans-Pritchard (1940, p. 63) summarizes the seasonal cycle as follows:

> The year thus consists of a period in villages and a period in camps, and the camp period is divided into the early part of the dry season when the young people move from small camp to small camp and the latter part of the dry season when everyone is concentrated in large camps around permanent water which they do not leave until they return to their villages.

In his summary of Nuer ecology Evans-Pritchard (1940, p. 92) presents an analysis which prefigures later work on human ecosystems. He argues that ecological relations appear to be in a state of equilibrium and that the balance between various resources in the context of environmental factors 'determines Nuer life at certain times of the year'. The hostile environment and the marginal nature of the economy produces 'interdependence between persons living in much larger areas and compels their acceptance of conventions of a political order' (Evans-Pritchard 1940, p. 93).

Evans-Pritchard correlates many other aspects of the Nuer way of life with the particular environmental circumstances in which they live. The presence of the cattle disease rinderpest, and adverse grazing conditions make it necessary to supplement cattle products with gardening. In the domain of culture, concepts of time are influenced by ecology, and Nuer aesthetics is the aesthetics of cattle (see Coote 1992). In the spheres of ritual and social relations, ceremonies and marriages are timed to coincide with the time of plenty that occurs at the transition between the dry season and the wet season. Even warfare is constrained by ecology since people lack the opportunity for war against the Dinka (a neighbouring group) at the height of the wet and the energy during the height of the dry. In all of these areas Evans-Pritchard is able to show individual correlations between ecology and a whole range of social and cultural variables, yet he never attempts a theoretical integration of the various strands. In his analysis there is almost a disconnection between the Nuer adaptive system and the political structure. The political system is treated as if it is a phenomenon *sui generis* that in its form is constrained by the environment and which is able to fulfil certain necessary tasks, but is not part of an integrated system.

> Physical environment, mode of livelihood, poor communications, a simple technology, and a sparse food supply—all in fact that we call their oecology—to some extent explain the demographic features of Nuer segmentation, but the

tendency towards segmentation must be defined as a fundamental principle of their social structure. (Evans-Pritchard 1940, p. 148.)

Evans-Pritchard's analysis of the Nuer shows a curious mixture of environmental determinism, environmental constraint, and social structural autonomy. This incoherence is the product of theoretical contradictions between a Radcliffe-Brownian functionalism, which reified social structure and almost endowed it with autonomous existence, and an ecological framework. It was one of the many contradictions that pervaded Evans-Pritchard's work, yet it is precisely his ability to write outside and between established theoretical frameworks and his refusal to be limited by a particular theory that make his writings such a rich source of ideas and ethnography. If the Nuer political system is necessary for adaptation to environmental circumstances (as Evans-Pritchard argues) then it is probable that characteristics of that system as an evolving form will be influenced by environmental factors and that aspects of the environment will be influenced by the particular form of the adaptive process. If the ecology is in a state of equilibrium then the social variables that are part of the maintenance of that state must be included within the analysis of the ecosystem and not bracketed off outside it. I will return again shortly to look at the Nuer from the perspective of a later analyst, Raymond Kelly, who adopts a more explicitly cultural interpretation of the dynamics of the system. Before doing so however I will take a detour via the New Guinea Highlands to look at an analysis that in many ways parallels that of Evans-Pritchard.

Pigs for the ancestors: integrating religion and environment

Roy Rappaport's *Pigs for the ancestors* (1967) was a landmark work in anthropology, since it was the first monograph to systematically apply the framework of ecology to develop an integrated explanation of all aspects of the socio-cultural system. Rappaport's study is based on the Tsembaga, a group of clans comprising some 200 Maring-speaking people who occupy an area of land on the south side of the Simbai Valley in the Highlands of Papua New Guinea. Rappaport carried out a detailed study of the subsistence economy of the Tsembaga in order to show the relationship between resources, labour, diet, and energy. The core of his analysis is focused on the pig population which in terms of calories only makes a minor (though in some contexts crucial) contribution to the diet. Pigs, however, have a major role in the overall eco-system in maintaining the balance between various resources and in structuring the cycle of the economy. Pigs are not a regular part of the diet, rather the herds are built up over a number of years until the time comes to hold one of the major regional ceremonies, the Kaiko. The Kaiko is an exchange ceremony that is also held to free the community from a series of taboos associated with warfare. In the years following a battle participants are forbidden from taking further part in warfare until they have held one of these regional ceremonies. And when one of the ceremonies is held the hosts will kill the larger proportion of their pig population.

Rappaport argues that the system is regulated overall by a series of feedbacks that ensure that each variable remains within certain limits that do not threaten other components of the system (1967, pp. 96–8). For example there is a clear relationship between the size of the pig population, the environment, and human labour. Pigs are left to obtain much of their diet from the forest and from human leftovers. However, their diet is supplemented with taro, which is also a part of the staple diet of the Tsembaga themselves. This supplementing of their diet helps keep the pig population under control both by attracting them back to the settlement and limiting their damage of the gardens. When the pig population is low then the system is not under any pressure. The pigs do not compete with humans for food but consume the surplus taro produced. They can even be seen to have a number of positive effects. The pigs can be thought of as storing surplus taro by converting it into meat (Rappaport 1967, p. 158). Moreover they help maintain the regional environment by allowing an area of forest to be kept free from cultivation, which is in effect returned to fallow when the pig population declines.

However, a point is reached where a number of variables exceed the margins of flexibility allowed by the system. The pig population will eventually reach a size when it begins to compete with humans for the staple crop, when it gets out of control and begins to destroy gardens, and when the labour required to care for the pigs becomes an intolerable burden on the women whose main responsibility they are. In fact when Rappaport was doing his fieldwork the key factor that operated to trigger the Kaiko and reduce the pig population was the increasing burden that was placed on individual women. 'The physiological capacity of women to care for pigs is below the capacity of the territory to provide the animals with sustenance' (Rappaport 1967, p. 162). Women complain to their husbands and brothers and eventually the pressure they bring to bear has its effect.

Rappaport discusses a number of different loops and feedback mechanisms which interrelate at various points to form a hierarchical structure which ensures that the Tsembaga operates as a self-regulating system. The highest level control mechanism is the Kaiko itself and the metaphysical beliefs in the nature of the spirit that underlie it. The Kaiko in turn is integrated into the regional system of warfare and exchange. The requirement to hold a Kaiko creates the economic cycles which are associated with the growth and destruction of the pig population. Although in each particular case a Kaiko is held on average perhaps every 10 years the fact that the cycles of different groups are out of phase with one another means that in the region as a whole such festivals take place on an annual basis. This ensures in effect the redistribution of high quality protein (though the overall nutritional significance of this is likely to be minimal) and provides the framework for regional political relations.

Rappaport's analysis has been criticized on a number of bases including problems with the detailed economic and ecological data (e.g. McArthur 1977), but the most recurrent theoretical critiques have been against the implicit functionalism of the model and the overemphasis on equilibrium (Friedman 1974). It is an oversimplification to represent Rappaport's analysis as presenting the Tsembaga as 'a homeostatic self-regulating system' (Kelly 1985). His analysis is however predicated on the assumption that in the given case the system has reached a stage of relative stability in the relationship between a set of critical variables. The logic of the position he adopts is that major changes in one or more of those variables will have consequences on other parts of the system that may result in system change or, in some circumstances, disaster. For example, an increase in the human population could result in famine, or an increase in the pig population beyond a certain point could result in the destruction of crops destined for humans or an intolerable increase in the burden of work. The assumption behind Rappaport's analysis is that over time human adaptive systems tend to develop a series of checks, balances, and interrelations among variables that ensures that in most circumstances the system can adjust itself so that disaster is avoided.

However, the type of analysis that Rappaport develops with all of its complex interconnections and multiple feedback loops does not preclude change, nor does he imply that the system is not in a dynamic and potentially evolving state. He argues that the Tsembaga population is well below its maximum and that population increase up to a certain level is certainly possible within the constraints of the system. He also shows how a change in certain variables such as selective weeding could increase the productivity of the system at certain points. However, he does give the impression that the system is a functionally interconnected whole that is to an extent overdetermined in the direction of equilibrium maintenance. The analysis has enabled Rappaport to demonstrate the logical interrelationship between certain variables and identify what some of the consequences of the alteration of particular variables might be, however the more those variables are linked into an overall structure of interdependence the more implausible the overall structure becomes and the less possible it becomes to connect it with processes that are operating on the ground.

In many ways the ground covered overlaps to an extraordinary extent with Evans-Pritchard's *Nuer*, but whereas Rappaport pushes the ecology as far as he can to explain features of Tsembaga society and integrates ecological variables within his explanation of religion and social organization, Evans-Pritchard erects boundaries that protect the autonomy of the social structure. For example in the section on ecology Evans-Pritchard (1940, p. 85) writes of the sharing of food that 'this mutual assistance and common consumption of food belongs rather to the subject of domestic and kinship relations than to the present account'. And later in his summary on political organization he concludes (Evans-Pritchard 1940, p. 190) 'ecology only helps us understand demographic features of Nuer tribes and tribal networks and not the nature of their structural relations'. Evans-Pritchard assumes categories whereas Rappaport dissolves them. Whereas Rappaport's system includes social and ecological variables, Evans-Pritchard uses the concept of system almost exclusively with reference to social and political organization. Yet the spirit of his analysis is much closer to Rappaport than the theoretical framework he uses implies. Indeed the stress on the separate status of the social structure can be interpreted as a sign of a remaining allegiance to the Radcliffe-Brownian functionalism that he was in the process of rejecting. In his initial statement of his aims he comes near to producing a paradigm statement of a sophisticated ecological anthropology.

> ... cattle are the dominant value in their [the Nuer] lives. We shall ask later how this value is related to environmental conditions and how far the two taken together, help us to explain the characteristics of Nuer political structure. (Evans-Pritchard 1940, p. 19)

This perspective, combined with Evans-Pritchard's conclusion that the Nuer system was in a state of equilibrium with the environment, explains why a later writer such as Kelly (1985) places the analyses of the Nuer by Evans-Pritchard and of the Tsembaga by Rappaport in the same box.

And back to the Nuer

Kelly in his reanalysis of the Nuer has argued that both Evans-Pritchard and Rappaport have overlooked the significance of cultural factors as determinants in the respective systems they analysed. Evans-Pritchard does so by holding certain components of the system separate from the rest, seeing, for example, such things as demography and political organization being acted upon by environmental constraints: 'harsh environment together with dominant pastoral interests cause low density' (1940, p. 263). Rappaport in contrast presents the system as an integrated whole, as a self-regulating system, but in his case it is equally as difficult to see how the component parts have come into existence. Religion, for example, is given a dominant role in governing the ecological cycle, yet the regulatory aspects of Tsembaga rituals are stated explicitly not to be part of people's conscious motivations (1967, p. 6). While many of the variables may have the effect that they are said to have and may be an integral part of the system as it operates at a particular point in time, their existence is to an extent independent from their integration within the system. The characteristics of the variables are in part culturally determined, though their expression is constrained by ecology and by their interaction with other variables. Even something as basic to survival as food, is both a component of an eco-system and a cultural value. A population's food requirements are not determined solely by physiological needs, but by those needs incorporated as part of a system of value (see Sahlins 1976). If a society is geared towards the production and consumption of an expensive resource (such as cattle in the case of the Nuer and pigs in the case of the Tsembaga) then as a variable that resource must reflect its cultural value rather than simply its value to survival. Likewise the means whereby resources are produced or regulated have a relatively autonomous existence as cultural forms independent of any function that they perform. Other institutions apart from the Kaiko could easily be imagined that would take on the self-regulating functions that it performs, a form of big man system, for example, as operates elsewhere in the Highlands.

While it is useful to model the interrelations between the variables that Rappaport identifies, and to demonstrate the consequences that are likely to follow from major alterations of particular variables, the causal factors involved still have to be identified. Clearly in the case of variables where change may have a catastrophic effect, negative feedback may be a sufficient explanation for maintaining the status quo, but systems do change all the time and self-regulation in itself cannot be the motive force. Indeed the wide range of values associated with each variable provides considerable leeway in the Tsembaga system which should create the space for change. The central role that Rappaport gives to the ritual system in the reproduction of Tsembaga society illustrates the problem of determinacy well—as a component of the system it plays a critical role in self-regulatory processes but it is one which people themselves do not recognize. The self-regulatory role cannot in itself be a sufficient explanation of the Kaiko even if it is a factor in its continuing existence and the way it articulates with other structures. If the Kaiko was not there then something else would have to be imagined in its place. Clearly that something else would be linked in different ways to other components of the socio-cultural system and we would be analysing a society with a different range of variables. The ritual system is not simply a component of a self-regulating system, but integral to the characteristics of that particular system. Self-regulation in itself is not the motivating factor behind the system, though it may explain the particular way another variable has been incorporated in the system and it may be part of the mechanism for change.

Kelly criticizes Evans-Pritchard's stable equilibrium model of the Nuer and goes on to argue that the historical trajectory of the Nuer has been determined as much by cultural as environmental factors. Kelly's reinterpretation of the Nuer data is based on two key facts which he argues are incompatible with the model of the Nuer as a system in equilibrium: one is that Nuer population density is less than that of the surrounding tribes, the other is that the history of the past three hundred years until *pax Brittanica* was established was a history of Nuer expansion into the territories of neighbouring groups who practised similar economies in similar environments with higher population densities. The greatest expansion occurred in the period between 1818 and 1890, when the Nuer increased their territory fourfold (Kelly 1985, p. 1). The main loss of land was suffered by the Dinka who speak a closely related language to the Nuer and who may indeed have been part of the same proto-population. The history of the relations between the Nuer and the Dinka provides Kelly with a case study of the 'interplay and relative significance of social and ecological factors' in accounting for territorial expansion —i.e. since the environment is a constant (though an extremely unpredictable constant!) it could be that one system of organization is better adapted to that environment or it could be that it is better adapted to warfare.

Kelly has three major objectives: one is to demonstrate the causal factors behind Nuer expansion, in other words why they needed extra land if their population density was lower than that of their neighbours; the second objective was to show how the Nuer were able to dominate numerically superior and equivalently armed neighbours; and the third objective was to show how the characteristics of the Nuer system that both motivated them to expand and provided them with the means to do so evolved out of a previously homogeneous system (Forde's 1970 analysis prefigures many of Kelly's arguments).

According to Kelly (1985, p. 112) 'the causal factors capable of providing a sustained impetus to Nuer expansion are those that affect the growth and demography of cattle'. Land shortage was created by inflationary processes inherent in the Nuer bridewealth system, which required that more cattle were locked up in a complex network of exchange among lineages in the case of the Nuer than the Dinka. Indeed he argues that in the case of the Dinka, in times of hardship and depreciated herds, marriages could take place with a token movement of cattle. In the case of the Nuer he argues that no such allowance was made and hence any reduction in the cattle population through disease or environmental disaster had to be compensated for by the seizure of cattle, land, or both. Once the group had moved into a new area then the need for increased land was maintained by inflation in marriage payments and by the fact that the majority of the conquered population was absorbed within the Nuer. The composition of marriage payments differed in the two cases. Nuer required oxen to be part of the bridewealth whereas in the case of the Dinka fewer oxen were required as marriage payments and many more were sacrificed in ritual. The consequence of this was to produce herds with different structures. In Nuer herds there was a reduction in milk production and an increase in the number of cattle required per capita. Thus the same area of land was able to support fewer people under Nuer occupancy than when peopled by Dinka. Because the Nuer absorbed most of the Dinka from the land they occupied, the system was not so much an expansion of the Nuer population as the replacement of one set of cultural traditions by another: 'an ecological succession of cultural forms' (Kelly 1985, p. 22).

The normal mechanism for expansion was a cattle raid followed by the establishment of a base in the new territory which was gradually consolidated. Kelly argues that it was the specific characteristics of the segmentary lineage system that gave the Nuer their advantage over the Dinka. The whole of Nuer society can be thought of as existing within an overarching framework comprising lineages and segments of lineages which are ordered in relation to one another on the basis of relative genealogical distance and proximity. Kelly argues that the particular form that the lineage organization takes enables groups to be mobilized rapidly across space, and action to be co-ordinated against distant segments of the Nuer or against outside groups.

> The rainy season maximised Nuer organizational advantage over the Dinka. During this period the Dinka were grouped into small clusters of communities isolated from their neighbours by long stretches of open savannah, in a seasonal mode of population distribution dictated by their economic adaptation to the environment. Although Nuer conform to the same requirements, their segmentary lineage organization enabled them to transcend environmentally determined disaggregation. In other words Nuer social organisation enabled men from a wide

range of communities to unite in large-scale raids at a time when seasonal economic exigencies dictated a disbursed distribution among those scattered settlements. (Kelly 1985, p. 47)

Kelly's argument diverges from Evans-Pritchard's most significantly in that he sees the basic demographic features of the system, the size of human population that the area can support and the structure of the cattle population generated in part by cultural factors. 'The distinctive features of Nuer economic organisation (*vis à vis* that of the Dinka) are grounded in the Nuer socio-cultural system and are not products of adaptation to the environment. The transhumant system of production shared by the two cultural groups does represent such an adaptation, but the Nuer and Dinka variants of this general system cannot be accounted for in adaptational terms.'

Kelly's arguments, like Rappaport's, can be criticized on a number of grounds. It can be argued that he defines the difference between the Nuer and the Dinka too neatly and that population movements are characteristic of a region with such a harsh environment characterized by periodic famine. Johnson (1989, p. 235) for example has argued that 'If Nilotic movements are looked at historically in something approaching the *longue duree* ... then it is possible to suggest that the Nuer are not so unique; that their recent expansion links them firmly with the other Western Nilotic societies which surround and have preceded them in the settlement of the Upper Nile floodplain.' Johnson (1986) sees the Nuer 'expansion' as a component of the region's flexibility in which groups expand and contract or amalgamate according to the exigencies of the environment and the uneven though regular occurrence of disaster in the form of flood, drought, and disease. However, such a perspective is quite complementary with Kelly's even though it does give adaption, in the form of the requirement of flexible and labile regional organization, an input into the regional system at a higher level. It is still necessary to explain why in the particular case the Nuer have a temporary advantage over adjacent groups.

On the one hand Kelly plays down the similarities between his analysis and Rappaport's and on the other underestimates the extent to which components of the socio-cultural system are influenced by adaptive processes. Kelly points out a number of feedback mechanisms and sets of interrelationships among variables. For example he demonstrates the interrelationship between bridewealth and lineage solidarity which in turn is related to the mechanism of Nuer expansion. Kelly is constructing an argument that is very similar to Rappaport's except that he sees the system in a state of expansion rather than equilibrium. He also fails to explore the extent to which the lineage system itself can be explained in part in adaptive terms, though he provides an excellent account of how it functions in that context. The lineage system which enables the Nuer to have an advantage over the Dinka also functions to organize the distribution of population and resources and enables the development of a flexible system of social control that can operate at different levels of group organization. Both of these features can be seen as functionally related to the transhumant system of production. The articulation of the lineage system with territorial organization enables the population to be efficiently organized in harmony with the seasonal cycle. As a corollary, it enables the function of social control to be exercised in a region of fluctuating group size. In separating out the transhumant system of production as adaptational while describing the lineage system as cultural Kelly is in danger of falling into the same error as Evans-Pritchard before him.

However, it would be wrong to imply that all components of a socio-cultural system are equally closely integrated with one another, or indeed that all must have some functional relationship to the adaptive system. While some societies may show considerable integration across component sub-systems such as religious, political, and economic institutions, others exhibit areas of autonomy or are characterized by sectors with contradictory or competing functions. Indeed it is quite possible that some sectors of a society may be central to the operation of the system of production and others may be relatively independent. It is difficult, however, to determine on a priori grounds, which sub-system is going to be linked to which others.

Lansing's (1991) detailed analysis of the ecology of Balinese irrigation systems provides an interesting example. Many previous analyses had assumed that control of the flow and distribution of waters through the irrigation system was organized by the political hierarchy, and may indeed have been a causal factor in the development of regional political structures (e.g. Wittfogel 1957). Lansing however was able to show that the control of the irrigation system was relatively independent from the political structure and was integrated within the religious system, water flow being co-ordinated through a linked structure of water temples. This system enabled irrigation agriculture to carry on quite independent of any changes in political boundaries or alterations in the political hierarchy as long as those changes left the authority of the temples intact. However, such apparently self-contained areas of integration within a system are only relative to their historical context. A changed world order and the creation of the Indonesian State meant that for a while the authority of the temples and the local decision makers was overridden in the interests of a new agricultural regime sponsored by the World Bank. The local control of the irrigation system had also acted as a means of co-ordinating action against pests. When this broke down it had consequences on production. A situation that was in this case eventually reversed when the problem was diagnosed and the temple system restored to use.

Regional adaptations and historical trajectories

On countless occasions people must have taken decisions which radically affected their adaptive relations with the environment influenced by ideologies and unaware of the long-term consequences of their decisions. The moves between hunting and gathering, subsistence agriculture, and pastoralism which have occurred throughout time in the East African cattle belt, have frequently involved people taking decisions that they have subsequently had to reverse. The move to pastoralism in the more marginal environments has frequently been followed by a return to hunting and gathering when drought and disease have killed the cattle (e.g. Hodder 1982). Such trajectories involving systematic reversals can be conceptualized as a single complex adaptive system that moves through different modes. Arguments that such cycles are in themselves a mode of adaptation are harder to sustain if, as seems to have been the case in East Africa, the long term consequences of the process has deleterious effects on the environment that eventually excludes some of the options.

Friedman's analysis of the alternation between different social and economic structures in Highland Burma based on Leach's work provides an example of the interaction between cultural evolution and environmental constraint. He shows how a particular sequence of types is produced as a component of an ecological cycle that is at least in part a consequence of the differential impact each type has on the environment. Leach identifies two main types of political structure in the region, one, *gumsa*, is a relatively hierarchical society with a complex internal ranking system. The other, *gumlao*, is relatively egalitarian. Both types of society share many features in common and the extreme forms represent the opposite ends of a continuum with a considerable range of variation between the two poles. Since the structure of *gumsa* society is maintained in part by payments to those in positions in the hierarchy it has built into it the requirements of a surplus together with certain inflationary tendencies. The consequence of this is to create pressure for increased production which results in increased stress for the producers, an increase in conflict with neighbouring groups, and eventually in degradation of the environment. At some stage in this process a *gumlao* revolt occurs in which the hierarchy is overturned and a more egalitarian regime develops which in turn reduces the level of production required. The transformed society, however, still retains the core structural properties of the old which in the long term, it is hypothesized, result in the repetition of the previous trajectory and the re-establishment of a more hierarchical form.

> In sum the dynamic of the Kachin system might be envisaged as an evolution towards increasing hierarchy and state formation which comes into contradiction with its own material constraints of reproduction but by way of *gumlao* revolts succeeds in re-establishing the conditions for a renewed evolution. (Friedman 1975, p. 186)

Underlying Friedman's analysis are two themes that are central to any discussion of the relationship between culture and adaptation. One is the relationship between a population and its resources, the other is the extent to which a cultural system can be seen to interact with an environment to produce a particular cultural form in an analogous relationship as that between genotype and phenotype. Two main models developed in population biology have been applied to the analysis of the relationship between human populations and their resource base. In one, a 'catastrophe' theory of population control, the population expands exponentially until it exceeds the resources available to sustain it. It then declines through the effects of disease or natural disaster to a much reduced level, and then the cycle begins again. The second model is one in which the population is sustained at a level that is below the maximum that the environment could support at its most abundant, by mechanisms that control the population at a stage where pressure is beginning to build up, without it ever having reached the point of catastrophe when the relationship between population and environment is out of control (see Carr-Saunders 1922; Peterson 1975). In the human case an additional variable that has to be added to the equation is 'given the particular way in which the resources are utilized' or 'given the particular techno-economic system'. Human populations can up to a point alter the balance of relationships in favour of sustaining an ever increasing human population by selectively increasing the productivity of the environment. Thus one form of human cultural adaptation is to change the conditions of the game by processes of intensification in association with technological change that can produce at one time the origins of agricultural societies and at another the Green Revolution. However such innovative responses are not routine features of socio-cultural evolution nor are they necessarily going to enable the population to avoid the catastrophe in all cases. It may well be that the limits of growth of a particular socio-economic trajectory still eventually result in the onset of a catastrophe and it may equally well be the case that in other circumstances, and if consciousness plays a part in the relationship between population and resources, a policy of no growth may be the option followed. Thus in any model of the relationship between human populations and the environment, culture must be built in as a component, both in the form of resources but also in the form of ideology and consciousness.

The role that culture has in Friedman's model is of an underlying organization of ideas which as well as containing the knowledge to practice a particular form of agriculture, contains a structure of possible ways of relating people to each other and to place. In different circumstances the core cultural structure produces more or less complex variants of the same social formation which subsequently develop along a particular pathway before collapsing in on themselves and beginning the cycle again.

In some respects Friedman's analysis is a temporal version of Sahlins' pioneering study of forms of social organization in Polynesia. Part of the difficulty of establishing if the particular socio-cultural formation is determined in part by some underlying core cultural structure is the difficulty of isolating that core structure from the complex and particularistic process of history of which it is a part. The history has no beginning and no end, and often the boundaries of the formation are little more than an arbitrary cut-off point in geographical space. In such circumstances how is it possible to locate this structure in the complexity of the data, without it being simply the imposition of one person's synchronous abstraction on a fluid, chaotic, complex, multiply determined, and unbounded process? In choosing to analyse Polynesian society Sahlins felt that he had found a test case in which at least some of the variables were controlled.

Sahlins' (1958, p. ix) aim was to 'study adaptive variation in culture' by relating 'differences in an aspect of the social systems ... to differences in the adaptation of cultures to their environments'. He chose Polynesian societies because they 'derive from a single source; they are members of a single cultural genus that has filled in and adapted to a variety of local habitats'. Polynesia consists of a large number of islands with very different environments which would be capable of sustaining very different densities of populations assuming they employed similar techno-economic systems. The islands were populated within a relatively short period of time by variants of the same founderculture, peoples who spoke similar languages, had similar technology, and had features of their social structure in common. Sahlins is able to argue from his analysis that variations in the form of the society, in particular the degree of stratification and the nature of kinship organization, can be correlated with the productivity of the economy, which in turn is linked with the natural resources and the physical geography of the respective islands. The interaction of geography with core culture does not account for all the differences, and clearly once the populations are established the particularities of their own history and circumstances are going to create potentially divergent trajectories in otherwise similar environments. Nonetheless Sahlins makes a fairly convincing case that techno-environmental factors can influence the way in which a culture develops as people move to occupy different areas of land, or, as in this case, move from island to island.

Subsequent studies of Polynesian societies have called some of Sahlins' analysis into question (see Friedman 1981; Hooper and Huntsman 1985). The overall picture that Sahlins presents is too simplified and does perhaps overemphasize the role of techno-environmental factors as determinants. Certainly in his later work he has moved away from this perspective to one which focuses more on the underlying structural similarities between different Polynesian cultures and the way in which cultural structure has been a component of historical process in Polynesia (see, for example, Hooper and Huntsman 1985; Sahlins 1986). Hooper (1985, p. 3) argues that 'Sahlins sought to account for the diversity by elements external to the inherent logic of the social and cultural systems with which [he was] concerned'. More recently Irwin (1992) has demonstrated how geographical distance and the frequency of inter-island contact is another component of the jigsaw and one that affected both rates of change and cultural affinity.

However, although a more complex picture has emerged in which more variables have to be taken into account and in which geography plays a lesser role, the idea that in Polynesia we can see the evolution and transformation of a particular cultural structure through its interaction with the environment retains its appeal. It is equally as naive to neglect the possible role of environmental factors and adaptive processes as it is to fail to recognize the relative autonomy of cultural factors. Sahlins' (1958) original analysis supports the idea that cultures are systems of organized information that create the worlds in which people live. The information provides a potential for action but it is not organized in a neutral way free from any cross-linkages. Its very organization creates particular constraints on that potential by incorporating it into a pre-existing structure. But external variables, such as environmental factors, become internal variables when human beings respond to them, and as such they affect the long-term transformation of the system.

It is possible that the processes that Friedman identifies in operation in Highland Burma may have an echo in Polynesia on islands where population pressures and environmental degradation result in the transformation of one form of social organization into another as a result of ecological change. The duration of the Polynesian system is not long enough for us to see the extent to which the system might have been one that regulated its relationship with the environment by transforming back on itself. It is equally possible that Polynesian societies could have been set on a course of divergence in which the societies at different points in the continuum became unrecognizable from the others. We are likewise lacking in data from the Indo-Burman region that convincingly show that Leach's and Friedman's hypotheses about the transformation from *gumsa* to *gumlao* and back reflect actual historical processes (see Nugent 1982) or reflect more a continuum of regional variants on a theme.

In both the Burmese and the Polynesian cases we are dealing with societies who are set on trajectories which, in favourable circumstances, result in the intensification of the system, in tendencies towards greater hierarchization and specialization of functions and statuses, and in an increase in food production and consumption. If nothing else intervenes, such as the development of different techno-economic complexes which enable an increase in food production or the development of homeostatic mechanisms, the societies are orientated towards catastrophe and to reconstruction at a different level of organization. However, there is evidence that some societies have maintained a much more stable relationship between population and resources over time by developing cultural structures and social practices that enable them to keep their populations from growing too large. In such cases instead of looking for cultural structures that cycle on a boom–bust basis, we are looking for structures that limit the trajectory they are on by reproducing the past in the present. Such structures may still be tested against environmental variation, since that is an independent variable that requires different responses to factors such as seasonality and differences in the biomass. The most comprehensive set of data on such a possible system is that on the Australian Aborigines.

Hunter–gatherer adaptations

Australia, like Polynesia but for different reasons, provides a laboratory for the study of human cultural processes. In the case of Polynesia the laboratory conditions were provided by the short duration of human colonization; in the case of Australia it is the long duration of a particular mode of life, hunting and gathering. Australia remained a continent of hunters and gatherers for a period of at least 50 000 years until the advent of the European colonists some 200 years ago (White and O'Connell 1982).

The model I will be developing is one that shows how cultural mechanisms played a role in maintaining a balance between population and resources over time. Before going on to analyse the Aboriginal adaptive system as it appears ethnographically, I will briefly examine the prehistoric context out of which it arises. Although the fact that Aboriginal society has existed in a state of equilibrium between population and resources is not strictly necessary to my argument, if it can be shown to have been a relatively stable system that reproduced its population and its structure over time, then that increases the likelihood that the adaptive components of the system were at least periodically used for that purpose. At this stage in our understanding of human cultural systems of adaptation, the development of understanding is going to be a complex and dialectical procedure. Models from the ethnographic present can be used to develop insight into prehistoric processes and encourage prehistorians to seek data that will confirm or disprove hypotheses, while in turn knowledge gained from trajectories abstracted from the past can be used to inform interpretations of the present (see Meehan and Jones 1988). Models that posit the existence of equilibriums, or at least the existence of relatively stable populations over time, are prime examples of ones that require a dialogue between prehistory and ethnography (c.f. Birdsell's various papers, 1953, 1968, 1970). As long as the data sets are kept separate and are not subject to reductionist selection, such a dialogue with prehistory can only be positive.

Although we have no means of determining the stability and growth of the Australian population over time and probably never will do (see, for example, Beaton 1990; Davidson 1990), it is nevertheless a reasonable hypothesis that at the time of European colonization the population was in a relatively stable equilibrium with the resource base. Whatever model of Aboriginal prehistoric demography is adopted, Beaton (1990, p. 32) argues that 'the population growth rate must have been very near stationary for long periods of time'. There is little evidence that people were starving, that society was pressured, that epidemics were endemic, or that land shortage was a cause of warfare between groups (though see Davidson 1990, p. 51, for references to exceptions on these accounts). On the other hand there is evidence that mechanisms of population control existed across Australia and that means existed to monitor the environment (Peterson 1972; Cowlishaw 1978). None of this means that Australia existed in a state of unchanging stability for 50 000 years. It is increasingly clear that the hunting and gathering activities of man coincided with a period of massive environmental change in Australia, associated with climatic change, but also associated with the development of a flora that was adapted to a regime of regular burning (Lewis 1986). That fire-regime vegetation favoured man both because it was compatible with hunting activities but also because of its richness in edible seeds and fruits. It may even have been that human action created the vegetational and environmental changes, not because they were consciously creating a long-term advantage for themselves, but because the cultural practice gave them an adaptive advantage which fed back into the ecosystem. Some climatologists are even boldly hypothesizing that Aboriginal practices have resulted in the extension of the monsoonal system to Northern Australia.

For much of their history the technology of the Aborigines has been characterized by its stability. Many of the major innovations, such as ground stone axes, happened early on (e.g. White 1967). But there is some evidence for an occasional revolution of the kind that is likely to have had consequences on demography and society, at least until its ramifications had worked their way through the system. The development of the small tool tradition with associated changes in material culture (more hafted implements), art styles (the production of vibrant effects), and language (the spread of Pama-Nyungan), that occurred some 5000 years ago (Bowdler and O'Connor 1991), was probably one of these revolutions. Indeed when the evidence from earlier periods becomes strong enough for us to make positive statements, it may turn out to have been the only one on such a scale.

The technological and cultural revolution of 5000 years ago spread very rapidly around Australia and it is reasonable to hypothesize that the Aboriginal societies reached a new balance between population and resources several millennia before European colonization. There were certainly continuing contacts with outside groups across the Torres Strait and the Timor Sea with Melanesians and Indonesians respectively (MacKnight 1976). Such contacts may have intensified and influenced the economies of Northern Australia somewhat in the past 500 years. Throughout Australia, moreover, there were trading networks and continual innovation in the areas of song, dance, myth, and in aspects of the form of material objects (Mulvaney 1976). However, the basic technological system has remained stable and it seems increasingly clear that the pattern of social organization that characterized Aboriginal society at European contact was established with the development and spread of the small tool tradition.

The cultural mode of Aboriginal adaptation

In recent years there has been considerable research on hunter–gatherer modes of adaptation (see Winterhalder and Smith 1981). The key topics have been the way in which a balance is maintained between the population and its resources (Peterson 1986) and the ways in which the economy is regulated (see Williams and Hunn 1986). Initial theories were based on relatively simple ecological hypotheses, that the nature of the resource base determined the mode of adaptation given a particular techno-economic system. Such things as the division of labour, the size and composition of the foraging group, and the seasonal pattern of movement, were thought to follow on almost naturally from the nature of the resource base and the seasonality of the environment (see, for example, some of the essays in Lee and DeVore 1968). As the 1970s progressed such resource-led models were increasingly seen to be inadequate since they failed to explain the variety of hunter–gatherer societies and the major differences in their institutional structures. While some societies such as the Hadza, the !Kung, the Inuit, and the Batek fitted these resource-based models at least on the surface reasonably well, others such as the Northwest coast Americans, the Calusa of Florida, and the Australian Aborigines did so less well. Woodburn (1982, 1988) in a series of influential articles applied the distinction between immediate and delayed return societies to distinguish between different kinds of hunter–gatherer societies. This distinction is based on the existence of qualitatively different strategies for harvesting and utilizing the resources of the environment. Immediate return societies tend to utilize their resources opportunistically, consuming their food within a maximum of a day or so of obtaining it. They take from the environment what they require to satisfy their immediate needs and leave the rest. In contrast delayed return societies tend to manage their resources by employing longer term strategies including in many cases the use of highly developed storage technologies. Although this is a useful conceptual distinction to make, as is the case with almost all simple dualisms, when it is applied as a typological device many societies do not easily fit into the categories created. Australian Aborigines show features of both types, and indeed different Aboriginal societies veer towards different ends of the continuum. However, Woodburn was struck by the fact that Aborigines are involved in long-term social processes in particular the organization of the relationships between men and women through complex systems of exchange. Although that is not the most convincing reason for classifying Aborigines as delayed return societies, Woodburn has identified a feature of Aboriginal society which is linked to long-term regulatory processes, even though in his analysis he never really develops the argument.

Although there is considerable variation on a number of dimensions between different Aboriginal societies across the continent, it is nonetheless possible to identify features that all have in common and that can be used to characterize an Aboriginal cultural structure in opposition, for example, to that of the San or the Inuit (c.f. Peterson 1979). I have argued elsewhere (Morphy 1988, 1990) that the core structure of eastern Arnhem Land societies consists of a particular way of conceptualizing sets of people through linkages to land-based ancestral beings in the context of a system of moiety organization. I will focus my analysis on the Arnhem Land case but will argue that many of the features can be seen to apply elsewhere in Australia, in particular the centrality of the creation of linkages between people, ancestral being, and place in the context of an ideology of spiritual continuity, that is validated by the intergenerational transfer of knowledge associated with place. The way in which such a structure articulates with factors such as dual organization or patrilineal organization varies greatly across the continent, though I would argue that in many such cases the differences can be explained in terms of environmental variables.

The key area that I will be focusing on is the way in which the core structure relates to the balance between population and resources over time. Most models of the organization of hunter–gatherer societies have to deal with the problem of the way in which the population can be organized into groups that enable it to efficiently utilize the resources of the environment over both the short term and the long term. It was originally a cliche that hunter–gatherer societies were organized on the basis of flexible band organization (see Lee and DeVore 1968). The band, the group that was involved in the daily round of hunting and gathering, had to be able to adjust its size and composition according to the task in hand, the abundance of the particular resource, the scarcity of the resource base at the particular time of year, and so on. In most environments, and in particular in the case of environments in which there was marked seasonal variation, the size and composition of the group would have to respond to contingent factors such as the availability of water, the abundance and duration of a particular resource, and so on. In Northeast Arnhem Land the size of the group of people living together at the same location will vary according to the time of year from an extended family to a group of two hundred. Neither of these groups is the band. In a sense, however, the band is the conceptual group in the middle of these two extremes of group size; it lies at the fulcrum of the flux in the population. On the ground it is the group (frequently a named group) associated with a particular range that it occupies for much of the year. The band in this sense is a normal and a normative group and numbers around 50–60 individuals; the extended families are what it breaks down to at certain times of the year when resources are widespread, the groups of 200 are groups that members of the band join at times of ceremonial activity or wet season retreat. However, not all members of the band will attend the same ceremonial gathering and the extended families that form, although smaller than the band, may contain people who had previously been with other bands. Flexible band organization is indeed an apt description of the changing membership of groups over time and of the variation in size of the living unit throughout the year, and it must be a component of any theoretical construction of the way in which people are distributed in groups on the ground in Arnhem Land. But it is very much a partial description since it says nothing about the mechanisms for organizing people into groups, nor does it take account of the reasons why people congregate together in groups. In some of the literature of the 1960s hunter–gatherer societies were written of as if the organization of people into groups was the result of some natural process, whereby the overall pattern of distribution was the result of adjustments at the level of individual behaviour. The pressure of population on local resources was relieved by people moving out to find more accommodating and friendlier life spaces. People were excluded from group membership or encouraged to move as the result of a deteriorating social environment or an increase in the work demanded.

Certainly pressure on resources is going to be a factor in motivating human action, in causing groups to split or individuals or groups to move on to fresher pastures. However, such factors usually operate in the context of sets of possible courses of action that vary across cultures precisely because there is a cultural dimension to the framing of human choice and determining human action. In a Yao village (Mitchell 1956) pressure on resources results in an increase of witchcraft accusations which may eventually result in village fission whereas in a Yanamamo village (Chagnon 1968) the precursor of village fission is likely to be an increase in intra-village violence. Possible courses of action are created by the cultural heritage in conjunction with ecological circumstances. Of course it is still possible to argue as some have done that the primary causal factor is the ecological imperative and that the cultural response can be seen to have resulted from the need to satisfy that requirement. The cultural form in this case has no determining existence independent of the material circumstances it helps to maintain. However, such a model too seems oversimplified, and neglects to take into account that basic point emphasized by Sahlins (1976) that culture may be as important in defining the goals of human action as it is in enabling people to achieve them. The fact that people are motivated to gather together in large numbers for the performance of religious rituals may be a more significant factor in determining the size of such gatherings than any contribution such large-sized gatherings may make to the efficient utilization of the resource base. Moreover, even though it is possible analytically to separate out different dimensions of the function of a particular institution or event, as a cultural reality those functions may not be separate to the people concerned. The ecological rationale for a large ceremonial gathering, even supposing it is recognized, may not be separable from the religious motivations of the participants (see

Lansing 1991, p. 129 for a comparable analysis of the role of the temple in Balinese irrigation agriculture). The religious motivations may be sufficient in themselves to mean that the adaptive system must be one that enables such events to take place. We are not concerned just with human survival but with the survival of particular ways of being human. Adaptive goals are cultural goals in as much as they include the survival of a society in continuity with existing, albeit continually transforming, systems of value and knowledge. In circumstances where survival under those conditions is no longer possible then the possibility of radical change exists, but the trajectory of that change is not determined by some abstract conceptualization of what is ecologically required, since it comes out of an already existing cultural structure in which any change is likely to be multiply determined and in which, for example, religious values still have to be taken into account as co-determinants. In human cultures something that at one level is enabling at another level is desirable in its own terms.

Peterson (1972, 1974, 1975, 1986) in a series of convincing articles has identified the main mechanisms and cultural structures that underpin the Australian system of flexible band organization and that enable both the short-term and long-term ordering of the relationship between people and resources in Arnhem Land. Peterson's key idea is that the system of clanship which exists in parallel with band organization provides a mechanism for creating long-term emotional attachments to place that ensure that certain people will provide the focal point of band organization. These same mechanisms ensure that there are people who are in a position to monitor the condition of the land to which they are emotionally attached and over which they exercise authority. The core of Peterson's (1972, p. 29) argument is that: 'Australian clan totemism would appear to be, then, a highly adaptive spatial classification of the population with the power to draw on and reinforce a fundamental aspect of human psychology, namely the tendency to form long-lasting sentimental attachments'. In order to generalize the idea throughout Australia it is necessary to replace the central place given to clan organization by the more general concept of a 'mechanism for defining groups and individuals in terms of their attachment to place'.

Peterson's (1972) arguments developed from his observation that although there appeared to be no necessary relationship between bands and clans, since the territory and membership of both was distinct, the average size of both groups was the same, around 50–60 individuals (though there was greater variation in the size of clans than there was in bands). Clans are exogamous patrilineal descent groups. The members of a clan are unlikely ever to meet together as a whole and only rarely will all members of a clan be together at the same large ceremonial gathering. The clan holds rights in sets of ceremonial objects and religious law, and, most importantly, holds rights to areas of land (Morphy 1990, 1991). Although rights in land do not involve exclusive utilization of its resources, permission from the owning clan is required by those who hunt and gather over it. Moreover, Peterson argues that there are boundary maintaining rituals that act in certain respects as a territorial defence mechanism (see also Williams 1986*a*).

The term band is used for the groups that actually use the land, and the area of land which they exploit is termed their range (the terminology derives from Stanner 1965). The range usually involves the territory of a number of clans and, conversely, clan territories, rather than being discrete areas of land, are divided into parcels which are distributed widely through the ranges of a number of bands. By definition, since clans are exogamous, the bands consist of men and women of clans of both moieties, and indeed many other factors result in band membership being drawn from a wide range of clans.

However, although clans and bands can be conceptualized as belonging to different levels of organization, the two interact to regulate people/environment relations. People's attachment and closeness to their clan territories varies over their lifetime and according to individualistic factors. Both women and men, as wives and sons-in-law, respectively, spend much of their young adult and early middle years living with relatives away from their own clan lands. However, although they do not spend the whole of their lives on or even close to their own clan lands Peterson (1972, 1974) argues that they remain emotionally attached to them and are continually involved in the political relations that centre on the clan's property including its land.

People are, in a sense, born into the landscape and as they grow older their relationship with the land is reinforced. Arnhem Landers believe in spirit conception, the existence of a spiritual entity that is necessary to conception. The ancestral beings created the form of the landscape and instituted the social and ceremonial practices that the human groups who succeeded them in the land continued to practice. They left behind them conception spirits that give rise to future generations. The conception spirits are associated with particular ancestral beings and specific areas of land. People are in a sense thought to come from the land of their conception spirit. In Arnhem Land this is nearly always from a place in the person's own clan lands, thus reinforcing the relationship between clan and individual identity. Even if the conception spirit comes from a place associated with a related clan of the same moiety, as does sometimes occur, the person's links with their own clan will be established through the performance of rituals associated with their own clan lands and through the acquisition of knowledge about them. On death it is believed that people's spirits will return to their own clan lands where they will join the reservoir of spiritual forces that becomes the future source of conception spirits. There is thus a cyclical relationship between human beings and the ancestral past that continually renews the relationship between people and social groups and place (see Morphy 1984). In addition this process establishes long lasting and deep emotional ties to particular places that are reinforced in the course of people's lives. Expressions of love of country and of obligations felt towards the land are a major feature of Aboriginal discourse.

Continually people will be drawn back to their own clan country for the performance of ceremonies, and even if they don't return physically, will maintain contact through the songs and sacred paintings belonging to their land that are performed at rituals held at a distance from it. Later on in life, particularly in the case of men but also sometimes in the case of women, people return to live permanently close to their clan lands. And on death, even if they are living away from their own land, people's spirits and sometimes their bones will be returned to the land from which they came, thus providing a focal point of continuity for future generations. The effect of this in organizing people/land relations is twofold: it provides a mechanism for organizing the relationship between people and place around people who have a long-term relationship with the land over which the band ranges, and it provides a means of monitoring the effect of population on the resources of the area (Peterson 1986, Chapter 6).

The majority of bands have resident one or two senior members of the owning clans of the territory over which they range. The leaders of the bands indeed tend to be drawn from this set of people, and in many respects the bands can be seen to form around a core of people who come from the owning clans. The band tends to consist of senior men of the group, some of their children, their spouses, affines, and other close and not so close relatives. The system of clan organization thus provides a series of fixed points around which the bands can be flexibly organized. The fact that bands and clans approximate each other in size means that emotional attachments to land which distribute people across the landscape also act as a spacing mechanism for band organization. Although movement between bands is fluid and although the criteria for band membership are multiple, the articulation of the band with the history of clans from the region and the importance of kinship relations centred on members of those clans, means that mechanisms exist for group formation and dispersal that create continuities over time.

The extent to which Arnhem Land society can be seen as a mechanism for reproducing a landscape divided up into cultural space for people to fill is indeed quite extraordinary. I have argued elsewhere (Morphy 1990) that the system of linking people, ancestral being, and place together has resulted in a very stable relationship between ancestral law and place and a very fluid relationship between actual groups of people and place. Clans are subject to a continuous process of expansion and decline, fission and fusion, just as bands come and go with the passage of the seasons. What remains more constant is the mythological division of space and the sets of songs, dances, paintings, and languages that are associated with those spaces and provide their substance. Although we cannot know in absolute terms how stable those mythspaces are, we do have evidence to show that in the short term they survive the redivision of space among social groups as alliances change and as one group succeeds another over time. The continuity is created by the enormous cultural effort invested in passing on and performing the mythological events that resulted in the creation of place, and in the fact that place becomes the idiom in terms of which other relationships are expressed. The fact that people are active agents in determining positions in space rather than inheritors of spatial positions that are in actual fact reproduced by the past (i.e. by descent), does not affect the stability of the spatial grid itself. I would not wish to argue that this cultural structuring of space is caused by the requirements of a hunter–gathering society to create an orderly relationship between people and space, but I would argue that the cultural system has co-evolved with that requirement and that certainly in Arnhem Land it is a positive component of the particular mode of adaptation.

The advantage that this system has over one whereby, for example, people were distributed over the landscape on an arbitrary basis (supposing such a system can be realistically conceptualized), is that it enables continuity over the generations and also that it locates authority with those who have maximum knowledge of the environment concerned. Arnhem Land systems of land tenure are oriented towards maintaining such continuities. The main way in which people demonstrate their rights to ownership of land is through showing knowledge of the mythological origins of the topography, which in turn of course requires knowledge of the topography itself. In order to get the mythological knowledge of the land, people have to spend much time with the custodians of the previous generation. As well as mythological law this knowledge also includes an understanding of the way the land was occupied in the past, who moved with whom and at what times of the year, and what resources were exploited (see Williams 1986*b*). The past history of human occupation of the land, often framed in terms of mythologically sanctioned rights (for example through the use of shared names), becomes the precedent for the future. The consequence of this is that the people around whom the bands coalesce are people with an immense knowledge of the land and are, in addition, people who, because of their age and religious authority, are able to have some influence over others.

Peterson (1975) argues that as well as a spacing mechanism this system can operate as a means of population control. Arnhem Landers did have the means to limit population through infanticide and abortion, and although the case is by no means conclusively proved, the system is one that at least has the potential to exercise some control on the relationship between population and resources. The decision on infanticide is likely to be an individual one which will be influenced by such factors as the individual's state of health, age, number of living children, and gender relations (see Cowlishaw 1978). But an awareness of environmental factors, through long-term association with an area of land and knowledge of its ecological history, could easily operate as a complementary component in decision making processes.[6] Women did give increased stress and effort associated with the food quest as a reason for infanticide and it is quite possible that the sanction of infanticide by senior men and women who are aware that the resources of the environment are under pressure, relative to some time in the past, may create a climate of opinion in which population control is encouraged for a time (for some data on infanticide and population control see Yengoyan 1972; Cowlishaw 1978; Reid and Gurruwiwi 1979). The existence of cultural and social mechanisms for monitoring the environment together with methods of population control, in conjunction with natural processes of fertility reduction that operate on human populations under conditions of stress and hunger, may have been sufficient to regulate the population. Population control need not have been a potential that was always utilized for it to have been a long-term factor in the adaptive system of Arnhem Land populations and one that influenced the cultural trajectory.

Although there is considerable variation in Aboriginal society across Australia, the core structure that I have described for Arnhem Land applies to many other cases. In particular most Aboriginal societies have mechanisms for linking people to place and ancestral being. However, there is some evidence that the form in which that relationship is established and the ways in which it articulates with other components of social organization, such as the clans and moieties, vary considerably. It would clearly be productive to see how such variations are correlated with variations in the environment and the prerequisites of the economies in context. While flexible band organization may be a requirement of hunter–gatherer economies in seasonal environments, the logistics and mechanics of organizing such flexibility will vary according to factors such as overall population density and the detailed pattern of resource distribution. While a detailed comparative analysis is not possible at this stage, a brief comparison with the Pintupi, a Western Desert group, proves productive (for details see Myers 1986*a*, 1986*b*).

The Pintupi live in a markedly different environment from that of Arnhem Land. The annual rainfall is sufficient to support only a very low population density. The seasonal factor in much of the Western Desert is not as a marked as elsewhere in Australia, though opportunistic response to rainfall when it occurs is an important strategy (Gould 1969). The low resource base, in combination with its relatively widespread and uniform distribution, means that the Pintupi used to move about in much smaller bands than the Arnhem Land people and that they had to move more frequently and over greater distances. In times of extreme water shortage the Pintupi were, however, forced to congregate in large groups around one of the few permanent water resources that existed in their country. The only other times at which large numbers congregated were when conditions permitted the formation of large ceremonial gatherings. The Pintupi range was larger than its Arnhem land equivalent and in many respects the bands existed in a much less structured relationship with the landscape. The pattern of movement and the formation of groups was largely contingent on the circumstances of the time. In such an environment it is not surprising that the groups or sets of people that became associated with sites in the land were more like the bands in their flexibility and in the composite nature of their membership than they were like clans. Indeed it would be impossible on demographic grounds alone for rights in land to be distributed on the basis of patrilineal affiliation to place as is the case in Arnhem Land, there are just far too many places and far too few people. (Of course it would be possible if people were organized into larger territorial groups but such was not the case.)

As in Arnhem Land, the Western Desert was cross-cut by the tracks of the ancestral beings on their journeys of creation. Although the influence of a site on the track does spread out into the surrounding area, the landscape is less accurately seen as being divided up into a chequer board pattern of ancestral territories or mythspaces than as an overlapping and interlocking network of sites and ancestral tracks. This fits in with the general pattern that Peterson (1986, p. 55, 56) identifies for the more arid parts of Central Australia (see also Layton 1983):

As one moves into the arid regions, estates become more clearly a cluster of points in space rather than bounded areas [O'Connell 1976, p. 5] ... Well defined boundaries reappear in the relatively rich hill country of the MacDonnell Ranges where the Aranda live ... But out in the desert proper, boundaries lose their significance and the focus is unequivocally on sites and the tracks that link them together.

The structure of the social group associated with place is also very different in the more arid zones. People become affiliated to segments of tracks by factors such as long-term association with the place, father or mother being associated with the place, it being the death place of father or mother, and so on. People need not be members of the same land holding group as their fathers and may change the main group that they are associated with during their lifetime. The group associated with a site may expand according to the influence that it has in the community, and the extension of group membership and the body of rightholders to a particular set of sites is integral to Pintupi political process.

Thus in the case of the Pintupi, people are attached to place as firmly as they are in Arnhem Land, but the groups associated with place are formed on a much more flexible basis. It is also more arguable that in the Desert the function of operating as a spacing mechanism for organizing the relationships between people and land is far less necessary. Although the desert is mapped out in ancestral terms, the evidence suggests that groups are not distributed evenly across the landscape, and totemic cult membership is not a key factor in structuring the relationship between people and the environment. If there is a functional component to the cultural core of Aboriginal society that I have identified, then it may not have applied equally everywhere. Indeed, the fact that in the Central Australian desert the landscape is mapped out in as much detail as elsewhere, and political life and individual identity still revolve around the identification of people with place, may be a sign that rather than the cultural structure being a component of the adaptation to the environment, the cultural structure is itself adapted to fit the environment.

It is reasonable to hypothesize that once the system of relating people to land through the medium of the Dreaming was established as a core component of an Aboriginal cultural structure, it was likely to spread as the population expanded. This situation could apply even in cases where people move into areas where the adaptive significance of a particular institution is neutral. The cultural system associated with the small tool tradition spread into the Western Desert after it had been consolidated in more fertile environments. In the Western Desert, with its low population density, with an environment that makes opportunism a necessity, territorial organization was fundamentally different from that of Arnhem Land and the better-watered environments. Yet the same structure of mythspaces was created and just as much effort was expended in creating attachments between people and land. However, the mechanism for attaching people to land was fundamentally different: Western Desert right-holding groups are more like Arnhem Land bands in their structure than they are like clans. The small population and the demographic near impossibility of clan organization of any depth meant that the linkage of people to land had to be on a different basis. The structure of the institution has been modified to allow the transfer of the core cultural value to a new environment in which it lacks the adaptive significance it had in the place where it evolved.

The relationship between cultural structure and long-term adaptive processes in Australia needs more detailed research before the full complexity of the historical process can emerge and the interaction between different factors—social, cultural, and environmental—can be established. With more data it may be possible to identify processes operating in the reverse direction to the ones outlined above. Cultural practices which developed in response to conditions in the more marginal environments may have subsequently spread to areas where they had no adaptive significance because they too became established as cultural practices with their own relatively autonomous existence and value. Some people have argued that the origin and spread of Australian section systems, which categorize kin into groups on a socio-centric basis, originated in marginal environments where they functioned to co-ordinate social relations on a regional basis and to facilitate marriage arrangements (see Yengoyan 1968). However McKnight (1981) has shown convincingly that the case is far from established. Part of the problem is that institutions and cultural structures exist independent of the particular contexts in which they occur and have a value independent of any adaptive function that they may have. This does not mean that the process of cultural transmission and change is uninfluenced by adaptive factors. However, it does mean that for any particular case it is extremely difficult to extricate adaptive factors from the complexities of historical processes which produced it, involving as they do a multiplicity of interacting determinants and motivational factors which exist at very different levels.

Conclusion

The logic of my last example is that even if something evolves as an adaptive system it doesn't necessarily transform as an adapted system when it moves into a new environment or when rapid environmental change occurs. Cultural structures have a coherence of their own that is relatively independent of any adaptive function, and hence have an impetus or a determining role in their own right. The Pintupi are created as a particular kind of Australian Aboriginal society, not as a version of the !Kung, even if the culture of the !Kung is equally if not better fitted to a desert environment. If a new environment results in certain aspects of the cultural structure being maladaptive, either those features will change or the move into the new environment will fail.

Identifiable failures through maladaption are likely to be rare in human populations: on the one hand many components of a cultural system will be adaptively neutral and on the other the relationship between environment and population is not a deterministic one, but one that allows many variants. Rhys Jones' (1977, 1978) analyses of the Tasmanian Aborigines show us that there is a great deal of freedom and flexibility for an economy in a moderate environment. Major changes in technology, social organization, and use of food resources can occur without significant effects on population size or viability. The existence of flexibility and of wide parameters for variation do not mean that the changes that do occur cannot be usefully analysed from the perspective of adaptation; rather their presence implies that adaptation is not in itself *the* determining factor. Adaptation in human populations cannot be reduced to a simple matter of life or death—the range of possible existences is so great. Indeed, human beings are on the whole generalized in their adaptation, by definition leaving themselves room to manoeuvre and having options to choose from. In particular human beings have the capacity to monitor their effect on the environment and hence to pre-adapt to negative change if it is foreseen.

In the case of hunting and gathering societies time is on the side of human populations. It is possible that the 'adaptive' practices of a particular hunter–gatherer trajectory may have long-term and irreversible consequences on the environment, as has been the case in Australia. It is also possible that in adverse environments individuals make foraging decisions that prove to be fatal and result in death from cold or thirst. On the whole, however, the processes of innovation and environmental change occur slowly enough for adjustments to be made and decisions to be reversed, and these adjustments may mask the role of adaptation in the system. Changes that are specifically adaptive are hard to identify in a cultural process characterized by cumulative changes, many of which are of no adaptive significance. This is especially so when those changes occur in an environment in which the parameters regulating the relationship between population and resources are broad.

It is perhaps ironic that hunting and gathering societies, with their relatively limited technologies, may have developed better medium-term institutional structures and ideologies for monitoring and responding to environmental change than most others in human history. With the development of agricultural societies the pace of change quickens, and in complex industrialized society it quickens still further, so that the possibility of adjustment becomes ever more problematical. It may be for such reasons that in pastoral societies and agricultural societies in marginal environments one often finds cyclical systems where socio-cultural trajectories never seem able to escape from the processes of their own destruction. Friedman's example from Highland Burma is one, the East African cattle belt another, and the Mountain Ok area of Papua New Guinea a third. In more favourable environments, societies escape from such cycles by a continual process of innovation punctuated by disaster, but again the process of change almost inevitably results in a speeding up of events that makes it difficult to act, or even to develop awareness of what is happening, in time. The level of consciousness required to monitor cultural processes and their interaction with the environment increases in its complexity with the speed and scale of the system.

In the case of hunter–gatherer societies it is possible to see how a system of environmental monitoring with mechanisms for response, that originally evolved on a local scale, could have consequences on a more regional scale. It is equally possible to see how similar systems could have developed in the case of other long-term adaptations to natural or humanly created environments, for example, in sedentary agricultural societies, as Lansing (1991) shows for Bali. In complex industrialized societies operating on an even more global scale it is harder to envisage such mechanisms. Even if it becomes possible to measure and accurately interpret environmental change on a global scale, the development of response mechanisms in a world divided into nation-states with conflicting national and local interests will be much harder to achieve. What unites hunting and gathering societies and modern industrialized societies is the human capacity for a consciousness that allows awareness of the environmental consequences of human action, and the imagination to think of ways of countering its effects.

Acknowledgements

I am grateful to Geoffrey Harrison for inviting me to write this paper and commenting on it, and for the stimulating environment of the Biosocial Society's workshop on human adaptation. I would like to thank Julia Harrison, Tim Ingold, Frances Morphy, Nic Peterson, and Veronica Strang for their helpful criticisms of the paper. This paper is part of my long-term research on human landscapes funded by the

Economic and Social Research Council. My research in Arnhem Land has been funded by the Australian Institute of Aboriginal and Torres Strait Islanders Studies.

Notes

1. Radcliffe-Brown (1952, p. 9) certainly recognized the way in which the concept of adaptation could be integrated within 'the scheme of interpretation of social systems'. But, as Ingold has pointed out to me, no-one really took up his ideas on the matter.
2. I am grateful to Tim Ingold for this formulation.
3. Nic Peterson has emphasized to me that Birdsell's argument is deterministic in the true sense, since he argues that rainfall *determines* the amount of plant cover which in turn *determines* the biomass of other animals and plants which in turn *determines* the number of people that can be supported by a given mode of adaptation. Rainfall as a measure is in itself too simple since it does not take into account the effects of factors such as temperature and relative rates of evaporation on the biomass. However the inclusion of such factors would not affect the principles behind the argument.
4. Nic Peterson reading this section argued that it would be preferable on most occasions to use the word influence rather than determine. 'If something determines something else the presence of one is not only necessary but sufficient for the other always to follow. Once you move away from that position then it is more the case of there being several influences some or all of which it may be possible to rank in significance.' While I see much sense behind the argument and agree that in many cases where people use the word determine a weaker word such as influence may be more appropriate, I do not think that the word determine needs to be restricted to cases where the factor is both necessary and sufficient. If things can be multiply determined and a set of factors can be isolated as co-determinants, why cannot x be something that determines y in conjunction with a–c in the context of d within the limits e? Although this uses a weaker concept of determinism than Peterson's I think that is probably desirable since it is moving away from reduction towards a more complex view of socio-cultural processes.
5. This dual aspect of the environment is captured well by Ingold (in press) when he writes of hunter–gatherers 'although the hunter–gatherer almost by definition may be supposed to inhabit a natural rather than an artificial environment, there is a sense in which the environment is nevertheless man-made. This 'making' consists in the mapping, onto the world of physical things, of a system of ideas or concepts regarding their future use, thereby converting those things—objectively presented to the organising subject—into use values. In other words making here refers to the cultural ordering of nature.'
6. I am grateful to Nic Peterson for this cautious formulation.

References

Beaton, J.M. (1990). The importance of past population for prehistory. In *Hunter–gatherer demography: past and present*, Oceania monograph 39 (ed. B. Meehan and N. White), pp. 23–40. Oceania Monographs, Sydney.

Birdsell, J.B. (1953). Some environmental and cultural facts influencing the structuring of Aboriginal populations. *Am. Naturalist*, **87**, 171–207.

Birdsell, J.B. (1968). Some predictions for the Pleistocene based on equilibrium systems among recent hunter–gatherers. In *Man the hunter* (ed. R. Lee and I. DeVore). Aldine, Chicago.

Birdsell, J.B. (1970). Local group composition among the Australian Aborigines: a critique of evidence from fieldwork conducted since 1930. *Curr. Anth.*, **11**, 115–41.

Bowdler, S. and O'Connor, S. (1991). The dating of the small tool tradition, with new evidence from the Kimberley, WA. *Aust. Aboriginal Stud.*, **1991** (1), 53–62.

Burnham, P. (1973). The explanatory value of the concept of adaptation in studies of culture change. In *The explanation of culture change: models in prehistory* (ed. C. Renfrew), pp. 93–102. Duckworth, London.

Burnham, P. (1979). Permissive ecology and structural conservatism in Gbaya society. In *Social and economic systems*, ASA Monograph 18 (ed. P. Burnham and R. Ellen). Academic Press, London.

Carr-Saunders, C.R. (1922). *The population problem: a study in human evolution.* Clarendon Press, Oxford.

Chagnon, N. (1968). *Yanamamo: the fierce people.* Holt Reinhart and Winston, New York.

Clark, C. and Haswell, M. (1971). *The economics of subsistence agriculture.* Macmillan, London.

Coote, J. (1992). 'Marvels of everyday vision': the anthropology of aesthetics and the cattle keeping Nilotes. In *Anthropology, art and aesthetics*, Oxford studies in the anthropology of cultural forms (ed. J. Coote and A. Shelton), pp. 245–73. Clarendon Press, Oxford.

Cowlishaw, G. (1978). Infanticide in Aboriginal Australia. *Oceania*, **48**, 262–83.

Davidson, I. (1990). Prehistoric Australian demography. In *Hunter–gatherer demography: past and present*, Oceania monograph 39 (ed. B. Meehan and N. White), pp. 41–58. Oceania Monographs, Sydney.

Dawkins, R. (1982). *The extended phenotype: the gene as a unit of selection.* Freeman, San Franciso.

Durham, W.H. (1991). *Coevolution: genes, culture and human diversity.* Stanford University Press.

Dyson-Hudson, N. (1966). *Karimojong politics.* Clarendon Press, Oxford.

Ellen, R.F. (1978). *Nuaulu settlement and ecology: an approach to the environmental relations of an Eastern Indonesian community.* (Verhandelingen van het Koninklijk Instituut voor Taal-, Land en Volkenkunde 83.) Martinus Nijhoff, The Hague.

Ellen, R.F. (1982). *Environment, subsistence and system: the ecology of small-scale social formations.* Cambridge University Press.

Evans-Pritchard, E.E. (1940). *The Nuer*. Clarendon Press, Oxford.
Forde, C.D. (1934). *Habitat, economy and society*. Methuen, London.
Forde, C.D. (1970). Ecology and social structure. *Proceedings of the Royal Anthropological Institute of Great Britain and Northern Ireland for 1970* pp. 15–29.
Friedman, J. (1974). Marxism, structuralism and vulgar materialism. *Man*, 9(3), 444–69.
Friedman, J. (1975). Tribes states and transformations. In *Marxist analyses and social anthropology* (ed. M. Bloch). Malaby Press, London.
Friedman, J. (1981). Notes on structure and history in Polynesia. *Folk*, 23, 275–95.
Godelier, M. (1977). *Perspectives in Marxist anthropology*. Cambridge University Press.
Gould, R.A. (1969). Subsistence behaviour among the Western Desert Aborigines of Australia, *Oceania*, 39(4), 253–74.
Harris, M. (1965). The myth of the sacred cow. In *Man, culture and animals: the role of animals in human ecological adjustments*, American Association for the Advancement of Science, Publication 78 (ed. A. Leeds and A.P. Vayda). American Association for the Advancement of Science, Washington, D.C.
Harris, M. (1979). *Cultural materialism: the struggle for a science of culture*. Harper Row, New York.
Harrison, G.A. (1972). Biological evolution and social change. *Proceedings of the Royal Anthropological Institute of Great Britain and Northern Ireland for 1971*, pp. 5–15.
Harrison, G.A. (1991). Ecological success and its measurement: introduction. *J. Hum. Ecol.*, Special Issue 1, 1–3.
Hodder, I. (1982). *Symbols in action: ethnoarchaeological studies of material culture*. Cambridge University Press.
Hooper, A. (1985). Introduction. In *Transformations of Polynesian culture* (ed. A. Hooper and J. Huntsman). The Polynesian Society, Auckland.
Hooper, A. and Huntsman, J. (1985). *Transformations of Polynesian culture*. The Polynesian Society, Auckland.
Ingold, T. (1980). *Hunters, pastoralists and ranchers*. Cambridge University Press.
Ingold, T. (1986). *Evolution and social life*, Themes in the social sciences. Cambridge University Press.
Ingold, T. (in press). Social relations, human ecology and the evolution of culture: an exploration of concepts and definitions. In *A Handbook of human symbolic evolution* (ed. A. Lock and C.R. Peters). Oxford University Press.
Irwin, G. (1992). *The prehistoric exploration and colonisation of the Pacific*. Cambridge University Press.
Johnson, D.H. (1986). The historical approach to the study of societies and their environment in the Eastern Upper Nile Plains. *Cahiers d'Etudes Africaines*, XXVI (1,2), 131–44.
Johnson, D.H. (1989). Enforcing separate identities in Southern Sudan: the case for the Nilotics of the Upper Nile. In *Les ethnies ont une histoire* (ed. J.-P. Cretien and G. Prunier), pp. 235–45. Karthala, Paris.
Jones, R. (1977). The Tasmanian paradox. In *Stone tools as cultural markers: change, evolution and complexity* (ed. R.V.S. Wright), pp. 189–204. Australian Institute of Aboriginal Studies, Canberra.

Jones, R. (1978). Why did the Tasmanians stop eating fish? In *Explorations in ethnoarchaeology* (ed. R.A. Gould), pp. 11–47. University of New Mexico Press, Albuquerque.
Kelly, R.C. (1985). *The Nuer conquest: the structure and development of an expansionist system*. University of Michigan Press, Ann Arbor.
Lansing, J.S. (1991). *Priests and programmers: technologies of power in the engineered landscape of Bali*. Princeton University Press.
Layton, R. (1983). Ambilineal descent and traditional Pitjantjatjara rights to land. In *Aborigines, land and land rights* (eds. N. Peterson and M. Langton), pp. 15–32. Australian Institute of Aboriginal Studies, Canberra.
Leach, E.R. (1954). *Political systems of Highland Burma*. Bell, London.
Lee, R. and DeVore I. (eds) (1968). *Man the hunter*. Aldine, Chicago.
Lewis, H.T. (1986). Fire technology and resource management in Aboriginal North America and Australia. In *Resource managers: North American and Australian hunter-gatherers* (ed. N.M. Williams and E. Hunn), pp. 45–67. Australian Institute of Aboriginal Studies, Canberra.
MacKnight, C.C. (1976). *The voyage to Marege: Macassan trepangers in northern Australia*. Melbourne University Press.
McArthur, M. (1977). Nutritional research in Melanesia: a second look at the Tsembaga. In *Subsistence and survival: rural ecology in the Pacific* (ed. T. Bayliss-Smith and R.G. Feachem). Academic Press, London.
McKnight, D. (1981). Distribution of Australian Aboriginal 'marriage classes'; environmental and demographic influences. *Man*, n.s. **16**, 75–89.
Meehan, B. and Jones, R. (eds) (1988). *Archaeology with ethnography: an Australian perspective*. Dept. of Prehistory, Research School of Pacific Studies, Canberra.
Meillasoux, C. (1981). *Maidens, meal and money*. Cambridge University Press.
Mitchell, J.C. (1956). *The Yao village*. Manchester University Press.
Moran, E.F. (1991). Human adaptive strategies in Amazonian Blackwater ecosystems. *Am. Anth.*, **93** (2), 361–82.
Morphy, H. (1984). *Journey to the crocodile's nest*. Australian Institute of Aboriginal Studies, Canberra.
Morphy H. (1988). Maintaining cosmic unity: ideology and the reproduction of Yolngu clans. In *Hunters and gatherers*, Vol. 2 *Property power and ideology* (ed. T. Ingold, J. Riches, and J. Woodburn), pp. 249–71. Berg, Oxford.
Morphy, H. (1990). Myth, totemism and the creation of clans. *Oceania*, **60**, 312–28.
Morphy, H. (1991). *Ancestral connections: art and an Aboriginal system of knowledge*. Chicago University Press.
Mulvaney, D.J. (1976). 'The chain of connection': the material evidence. In *Tribes and boundaries in Australia* (ed. N. Peterson), pp. 72–94. Australian Institute of Aboriginal Studies, Canberra.
Myers, F.R. (1986*a*). Always ask: resource use and land ownership among the Pintupi Aborigines. In *Resource managers: North American and Australian hunter-gatherers* (ed. N.M. Williams and E. Hunn), pp. 173–95. Australian Institute of Aboriginal Studies, Canberra.
Myers, F.R. (1986*b*). *Pintupi country, Pintupi self: sentiment, place and politics among Western Desert Aborigines*. Smithsonian Institution Press, Washington.

Nugent, D. (1982). Closed systems and contradiction: the Kachin in and out of history, *Man*, **17** (3), 508–27.
O'Connell, J.F. (1976). *Report on investigation of Alyawarra land claims*. Unpublished paper. Central Land Council, Alice Springs.
Peterson, N. (1972). Totemism yesterday: sentiment and local organisation among the Australian Aborigines. *Man*, **7** (1), 12–30.
Peterson, N. (1974). The importance of women in determining the composition of residential groups in Aboriginal Australia. In *Women's role in Aboriginal society* (ed. F. Gale), pp. 17–27. Australian Institute of Aboriginal Studies, Canberra.
Peterson, N. (1975). Hunter–gatherer territoriality: the perspective from Australia. *Am. Anth.*, **77**, 53–68.
Peterson, N. (1979). Territorial adaptations among desert hunter–gatherers: the !Kung and Australians compared. In *Social and ecological systems* (ed. P. Burnham and R. Ellen), pp. 111–29. Academic Press, London.
Peterson. N. (1986). *Australian Aboriginal territorial organisation*, Oceania Monograph 30. Oceania Monographs, Sydney.
Radcliffe-Brown, A.R. (1952). *Structure and function in primitive society*. Cohen and West, London.
Rappaport, R.A. (1967). *Pigs for the ancestors: ritual in the ecology of a New Guinea people*. Yale University Press, New Haven.
Reid, J. and Gurruwiwi, M. (1979). Attitudes towards family planning among women of a Northern Australian Aboriginal community. *Med. J. Aust., Special Suppl.*, **1**, 5–7.
Richards, A.I. (1939). *Land, labour and diet in Northern Rhodesia*. Routledge, Keegan Paul, London.
Sahlins, M. (1958). *Social stratification in Polynesia*. University of Washington Press, Seattle.
Sahlins, M. (1976). *Culture and practical reason*. Chicago University Press.
Sahlins, M. (1986). *Islands of history*. Chicago University Press.
Smith, E.A. (1983). Anthropological applications of optimal foraging theory: a critical review. *Curr. Anth.*, **24**, 625–51.
Smith, E.A. (1988). Risk and uncertainty in the 'original affluent society': evolutionary ecology of resource sharing and land tenure. In *Hunters and gatherers*, Vol. 1 *History, evolution and social change* (ed. T. Ingold, J. Riches, and J. Woodburn), pp. 222–51. Berg, Oxford.
Sponsel, L.E. (1986). Amazon ecology and adaptation. *Annu. Rev. Anth.*, **15**, 67–97.
Stanner, W.E.H. (1965). Aboriginal territorial organisation: estate, range, domaine and regime. *Oceania*, **36** (1), 1–26.
Steward, J. (1955). *The theory of culture change*. The University of Illinois Press, Urbana.
Thomson, D.F. (1939). The seasonal factor in human culture. *Proc. Prehistoric Soc.*, **5**, 209–21.
Wallace, A.R. (1870). *Contributions to the theory of natural selection*. Macmillan, London.
White, C. (1967). Early stone axes in Arnhem Land. *Antiquity* **41**, 149–52.
White, J.P. and O'Connell, J.F. (1982). *A prehistory of Australia, New Guinea and Sahul*. Academic Press, Sydney.

Williams, N.M. (1986*a*). A boundary is to cross: observations on Yolngu boundaries and permission. In *Resource managers: North American and Australian hunter–gatherers* (ed. N.M. Williams and E. Hunn), pp. 131–53. Australian Institute of Aboriginal Studies, Canberra.
Williams, N.M. (1986*b*). *The Yolngu and their land.* Australian Institute of Aboriginal Studies, Canberra.
Williams, N.M. and Hunn, E. (1986). *Resource managers: North American and Australian hunter–gatherers.* Australian Institute of Aboriginal Studies, Canberra.
Winterhalder, B. and Smith, E.A. (eds) (1981). *Hunter–gatherer foraging strategies: ethnographic and archaeological analyses.* Chicago University Press.
Wittfogel, K.A. (1957). *Oriental despotism: a comparative study of total power.* Yale University Press, New Haven.
Woodburn, J. (1968). Stability and flexibility in Hadza residential groupings. In *Man the hunter* (ed. R. Lee and I. DeVore). Aldine, Chicago.
Woodburn, J. (1982). Egalitarian societies. *Man*, 17, 431–51.
Woodburn, J. (1988). African hunter–gatherer social organisation: is it best understood as a product of encapsulation. In *Hunters and gatherers*, Vol. 1 *History, evolution and social change* (ed. T. Ingold, J. Riches, and J. Woodburn), pp. 31–64. Berg, Oxford.
Yengoyan, A. (1968). Demographic and ecological influences on Aboriginal Australian marriage sections. In *Man the hunter* (ed. R. Lee and I. DeVore). Aldine, Chicago.
Yengoyan, A. (1972). Biological and demographic components in Aboriginal Australian socio-economic organisation. *Oceania*, 43, 85.

INDEX

Lightning Source UK Ltd.
Milton Keynes UK
176339UK00002B/1/P

9 781859 739587